The Biology of Aging

REGENERATIVE MEDICINE

EPIGENETICS

IMMUNE SYSTEM

STEM CELLS

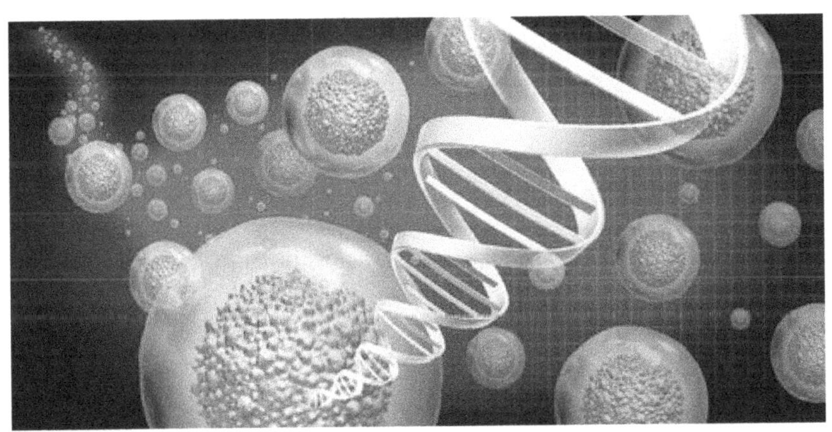

by

Enzo Milano

Table of Contents

KEY TOPICS

The biology of aging is a complex and multifaceted process that involves the gradual decline in function of various bodily systems over time. This decline is driven by a combination of genetic and environmental factors, including telomere shortening, DNA damage, epigenetic changes, loss of stem cell function, mitochondrial dysfunction, inflammation, hormonal changes, glycation, autophagy, sarcopenia, osteoporosis, and cognitive decline.

Telomeres are the protective caps on the ends of chromosomes that shorten with each cell division. When telomeres become too short, the cell can no longer divide and will enter a state of senescence or undergo programmed cell death, also known as apoptosis. Telomere shortening is a natural part of the aging process, but it can be accelerated by stress, smoking, and other environmental factors.

DNA damage can also contribute to the aging process. Our DNA is constantly being damaged by environmental factors such as UV radiation, oxidative stress, and errors during DNA replication. This damage can lead to mutations in genes that regulate cellular processes, contributing to aging.

Epigenetic changes refer to chemical modifications to DNA or histone proteins that can affect gene expression without altering the underlying DNA sequence. These changes can also contribute to aging by disrupting normal cellular processes.

Loss of stem cell function is another important factor in the aging process. Stem cells are the body's master cells, capable of differentiating into various cell types to repair and maintain tissues. With age, stem cell function declines, leading to a decrease in tissue regeneration and an increase in aging-related diseases.

Mitochondrial dysfunction is also a key player in the aging process. Mitochondria are the powerhouses of cells, responsible for generating energy in the form of ATP. Mitochondrial dysfunction can lead to the production of reactive oxygen species (ROS), which can damage cellular components and contribute to aging.

Inflammation is another major contributor to the aging process. Chronic, low-grade inflammation can occur with age and contribute to aging-related diseases such as arthritis, cardiovascular disease, and dementia.

Hormonal changes can also play a role in the aging process. Hormones such as estrogen and testosterone decline with age, leading to changes in various bodily systems and contributing to aging-related changes.

Glycation is the non-enzymatic attachment of sugars to proteins or lipids, leading to the formation of advanced glycosylation end-products (AGEs). AGEs can accumulate over time and contribute to aging-related diseases such as diabetes, cardiovascular disease, and Alzheimer's disease.

Autophagy is the cellular recycling process that helps maintain cellular homeostasis by removing damaged cellular components. Autophagy can decline with age, leading to an accumulation of damaged cellular components and contributing to aging-related diseases.

Sarcopenia is the loss of muscle mass and strength that occurs with age, leading to decreased mobility, frailty, and an increased risk of falls and fractures.

Osteoporosis is the thinning and weakening of bones that occurs with age, leading to an increased risk of fractures.

Cognitive decline is a natural part of the aging process, but it can be accelerated by various factors such as genetics, lifestyle, and environmental factors.

In conclusion, the biology of aging is a complex and multifaceted process that involves the gradual decline in function of various bodily systems over time. Understanding the biology of aging is crucial for developing effective strategies for healthy aging and preventing aging-related diseases.

Introduction

The biology of aging is a complex and multifaceted process that involves the gradual decline in function of various bodily systems over time. Here are some key biological factors that contribute to the aging process:

- Telomeres: Telomeres are the protective caps on the ends of chromosomes that shorten with each cell division. When telomeres become too short, the cell can no longer divide and will enter a state of senescence or undergo programmed cell death, also known as apoptosis.

- DNA damage: DNA damage can accumulate over time due to environmental factors such as UV radiation, oxidative stress, and errors during DNA replication. This can lead to mutations in genes that regulate cellular processes, contributing to aging.

- Epigenetic changes: Epigenetic changes refer to chemical modifications to DNA or histone proteins that can affect gene expression without altering the underlying DNA sequence. These changes can also contribute to aging by disrupting normal cellular processes.

- Loss of stem cell function: Stem cells are the body's master cells, capable of differentiating into various cell types to repair and maintain tissues. With age, stem cell function declines, leading to a decrease in tissue regeneration and an increase in aging-related diseases.

- Mitochondrial dysfunction: Mitochondria are the powerhouses of cells, responsible for generating energy in the form of ATP.

Mitochondrial dysfunction can lead to the production of reactive oxygen species (ROS), which can damage cellular components and contribute to aging.

- Inflammaging: Inflammaging refers to the chronic, low-grade inflammation that occurs with age and contributes to aging-related diseases such as arthritis, cardiovascular disease, and dementia.

- Hormonal changes: Hormonal changes can occur with age, particularly the decline in hormones such as estrogen and testosterone, which can impact various bodily systems and contribute to aging-related changes.

- Glycation: Glycation is the non-enzymatic attachment of sugars to proteins or lipids, leading to the formation of advanced glycosylation end-products (AGEs). AGEs can accumulate over time and contribute to aging-related diseases such as diabetes, cardiovascular disease, and Alzheimer's disease.

- Autophagy: Autophagy is the cellular recycling process that helps maintain cellular homeostasis by removing damaged cellular components. Autophagy can decline with age, leading to an accumulation of damaged cellular components and contributing to aging-related diseases.

- Sarcopenia: Sarcopenia is the loss of muscle mass and strength that occurs with age, leading to decreased mobility, frailty, and an increased risk of falls and fractures.

- Osteoporosis: Osteoporosis is the thinning and weakening of bones that occurs with age, leading to an increased risk of fractures.

- Cognitive decline: Cognitive decline is a natural part of aging, but it can be accelerated by various factors such as genetics, lifestyle, and environmental factors.

These biological factors can interact with each other and with environmental and lifestyle factors to contribute to the aging process. Understanding these factors can help us develop strategies to promote healthy aging and prevent aging-related diseases.

Indeed, our feelings about aging are complex and multifaceted. On the one hand, we admire and respect elderly individuals who remain active, healthy, and engaged in their communities. We celebrate their longevity and achievements, and we often look to them for wisdom and guidance. On the other hand, we also fear aging and the physical, cognitive, and emotional changes that come with it. We worry about losing our youth, energy, and independence, and we fear the stigma and marginalization that can accompany old age.

This ambivalence towards aging is reflected in our cultural attitudes and beliefs. On the one hand, we value longevity and seek to extend life through medical advances and healthy lifestyles. We celebrate birthdays and milestone events, and we often use age as a measure of accomplishment and achievement. On the other hand, we also perpetuate negative stereotypes about aging, portraying older adults as weak, frail, and dependent. We often dismiss or ignore the contributions and experiences of elderly individuals, and we may even fear or resent their presence in our communities.

So, what can we do to reconcile these conflicting attitudes towards aging? Firstly, we need to recognize and challenge our own biases and stereotypes about aging. We need to acknowledge the value and worth of all individuals, regardless of their age or abilities. We can do this by promoting intergenerational dialogue and collaboration, and by celebrating the achievements and contributions of older adults.

Secondly, we need to work towards creating a society that values and supports aging. This means investing in healthcare, social support, and community resources that enable older adults to live healthy, engaged, and fulfilling lives. We need to create age-friendly communities that are designed for people of all ages, with accessible housing, transportation, and social spaces.

Finally, we need to embrace aging as a natural and inevitable part of life. We need to recognize that aging is not something to be feared or denied, but rather something to be embraced and celebrated. We can do this by promoting positive representations of aging in our culture, and by fostering a sense of purpose and meaning in our lives, regardless of our age.

Our feelings about aging are complex and multifaceted, reflecting both our admiration for elderly individuals and our fears about the aging process. By recognizing and challenging our biases, creating age-friendly communities, and embracing aging as a natural part of life, we can work towards a society that values and supports aging, and where all individuals can live healthy, engaged, and fulfilling lives, regardless of their age.

AGING UNDER THE MICROSCOPE

It's true that many of us hope to live a long and healthy life, but the reality is that aging is a natural part of life, and it's important to be prepared for the changes that come with it. While it's understandable to want to stay young and vigorous, it's important to recognize that aging is a natural part of life, and it's important to take care of ourselves as we age.

It's important to remember that aging is not just about physical changes, but also about mental and emotional changes. As we age, it's important to stay engaged, connected, and active, both physically and mentally. This can help to maintain cognitive function, reduce the risk of depression and anxiety, and improve quality of life.

It's also important to recognize that aging is not something to be feared or denied, but rather something to be embraced and celebrated. Every stage of life has its own unique joys and challenges, and aging is no exception. By embracing aging and taking care of ourselves, we can live a fulfilling and meaningful life, even as we age.

In terms of day-to-day choices, it's important to make healthy lifestyle choices, such as eating a balanced diet, getting regular exercise, getting enough sleep, and managing stress. These choices can help to maintain physical and mental health, and reduce the risk of chronic diseases such as heart disease, diabetes, and dementia.

It's also important to stay engaged and connected with others, whether through work, volunteering, or social activities. Social isolation can have negative effects on mental and physical health, and it's important to stay connected with others to maintain a sense of purpose and belonging.

It's important to embrace aging and take care of ourselves, both physically and mentally, in order to live a fulfilling and meaningful life. By making healthy lifestyle choices and staying engaged and connected, we can live a long and healthy life, and make the most of every stage of life.

The topic of aging is a complex and multifaceted one, with many different factors contributing to the aging process. One of the most significant factors is the accumulation of cellular damage, which can occur as a result of a variety of factors such as environmental stressors, errors during cellular division, and the natural decline in cellular function that occurs with time. This damage can lead to a decrease in the function of our cells, tissues, and organs, which can result in the outward signs of aging that we see, such as wrinkles, gray hair, and decreased energy levels.

Another key factor in the aging process is inflammation, which can arise from a variety of sources, including the accumulation of cellular debris, the activation of pro-inflammatory genes, and the dysregulation of our immune systems. Chronic, low-grade inflammation can lead to the degradation of our tissues and organs, further contributing to the aging process.

Hormonal changes also play a significant role in the aging process. As we age, our hormone levels can become imbalanced, leading to a range of symptoms, including decreased muscle mass, bone density, and cognitive function. For example, the decline in growth hormone and estrogen levels during menopause can lead to a decrease in bone density and an increase in body fat.

In addition to these molecular and cellular changes, aging is also influenced by our lifestyle choices and environmental factors. Smoking, lack of exercise, and poor diet can all contribute to the aging process by increasing oxidative stress, inflammation, and cellular damage.

So, what can we do to slow down the aging process? While there is no fountain of youth, there are a few strategies that have been shown to slow down the ticking clock. Regular exercise, for example, can increase our cells' energy production, improve our insulin sensitivity, and reduce inflammation. A healthy diet rich in fruits, vegetables, whole grains, and healthy fats can help to reduce oxidative stress and inflammation, while also providing our bodies with the nutrients we need to function optimally. Getting enough sleep is also crucial, as it allows our bodies to repair and regenerate our cells.

Managing stress is also essential for slowing down the aging process. Chronic stress can lead to an increase in inflammation and oxidative stress, as well as a decrease in our immune function. Engaging in stress-reducing activities such as meditation, yoga, or deep breathing can help to mitigate the negative effects of stress and promote well-being.

Aging is a complex process that is influenced by a variety of molecular, cellular, and hormonal changes, as well as our lifestyle choices and environmental factors. While there is no fountain of youth, there are a few strategies that have been shown to slow down the aging process. By taking care of our bodies and minds, we can help to maintain our health and well-being as we age.

Indeed, the field of gerontology has made significant progress in recent years, thanks to advances in technology and collaboration among scientists. The National Institute on Aging (NIA) has played a crucial role in supporting and coordinating research on aging since its establishment in 197 - With the help of cutting-edge technologies, such as genomics, proteomics, and imaging techniques, researchers have been able to study the aging process at the molecular, cellular, and organismal levels.

One of the most significant breakthroughs in aging research has been the discovery of telomeres, the protective caps at the end of chromosomes that shorten with each cell division. Scientists have found that telomere shortening is a hallmark of aging and that it can lead to cellular senescence, a state in which cells are no longer able to divide and function properly. This discovery has led to a greater understanding of the role of telomeres in aging and the development of age-related diseases.

Another area of significant progress has been the study of senolytic therapy, which aims to remove senescent cells that accumulate with age and contribute to age-related diseases. Researchers have identified several senolytic drugs that can selectively eliminate senescent cells, improving healthspan and extending lifespan in animal models. Clinical trials are now underway to test the safety and efficacy of these drugs in humans.

The study of epigenetics, which examines how environmental factors can affect gene expression, has also shed light on the aging process. Researchers have found that aging is associated with changes in epigenetic marks, such as DNA methylation and histone modifications, which can affect the expression of genes involved in cellular maintenance and repair.

In addition, advances in stem cell biology have allowed scientists to study the aging of stem cells, which are critical for tissue maintenance and regeneration. Researchers have found that stem cells exhibit a decline in function with age, leading to a decrease in tissue repair and regeneration. This decline can contribute to the development of age-related diseases, such as osteoarthritis and atherosclerosis.

The use of computational models and machine learning algorithms has also been instrumental in aging research. These tools allow scientists to analyze large datasets and simulate complex biological processes, enabling them to identify patterns and relationships that would be difficult to discern through experimental methods alone.

Finally, the NIA's Interventions Testing Program (ITP) has played a crucial role in identifying potential interventions that could slow, halt, or even reverse the aging process. The ITP has supported clinical trials of several promising interventions, including metformin, rapamycin, and NAD+ supplementation.

The field of gerontology has made significant progress in recent years, thanks to advances in technology and collaboration among scientists. The discovery of telomeres, the study of senolytic therapy, the examination of epigenetics, the study of stem cell aging, and the use of computational models and machine learning algorithms have all contributed to a greater understanding of the aging process. The NIA's Interventions Testing Program has also played a crucial role in identifying potential interventions that could slow, halt, or even reverse the aging process. With continued research and collaboration, scientists are poised to make even greater strides in understanding and addressing the challenges of aging.

The mysteries of aging have fascinated scientists and philosophers for centuries, and despite significant progress, the underlying mechanisms of aging remain poorly understood. Theories about aging have come and gone, but a comprehensive explanation for the aging process is still elusive.

Aging is an inevitable part of life, and it's a complex process that affects individuals in different ways. While some people age gracefully, others experience a decline in physical and cognitive abilities. The aging process is characterized by a range of biochemical, genetic, and physiological changes that are not yet fully understood.

One of the most significant challenges in understanding aging is the sheer complexity of the process. Aging is influenced by a variety of factors, including genetics, lifestyle, and environmental factors, which interact in complex ways. Additionally, aging is a dynamic process that involves the gradual decline of various physiological systems, including the cardiovascular, immune, and nervous systems.

Despite these challenges, researchers have made significant progress in recent years. Advances in molecular biology, genetics, and imaging technologies have allowed scientists to study the aging process at the cellular and molecular level. For example, researchers have identified key players in the aging process, such as telomeres, senescent cells, and epigenetic changes.

Telomeres are the protective caps at the end of chromosomes that shorten with each cell division. As telomeres shorten, cells may reach a point where they can no longer divide, leading to cellular senescence. Senescent cells can contribute to the aging process by releasing pro-inflammatory cytokines and other factors that can damage surrounding tissues.

Epigenetic changes, such as DNA methylation and histone modifications, can also contribute to the aging process. These changes can affect the expression of genes involved in cellular maintenance and repair, leading to a decline in tissue function.

In addition to these molecular changes, the aging process is also influenced by genetic and environmental factors. Genetic mutations can affect the function of genes involved in the aging process, while environmental factors, such as stress, smoking, and poor diet, can accelerate the aging process.

While the facts of aging are still not fully understood, researchers are making progress in uncovering the underlying mechanisms. Advances in technology, such as single-cell analysis and machine learning algorithms, are allowing scientists to study the aging process in greater detail than ever before.

The aging process is a complex and multifaceted phenomenon that is still poorly understood. While theories about aging have come and gone, researchers are making progress in uncovering the underlying mechanisms of aging. By studying the aging process at the molecular, genetic, and physiological levels, scientists are poised to make significant breakthroughs in our understanding of aging and, ultimately, to develop interventions that can promote healthy aging and improve quality of life for older adults.

POSSIBLE PATHWAYS LEADING TO AGING

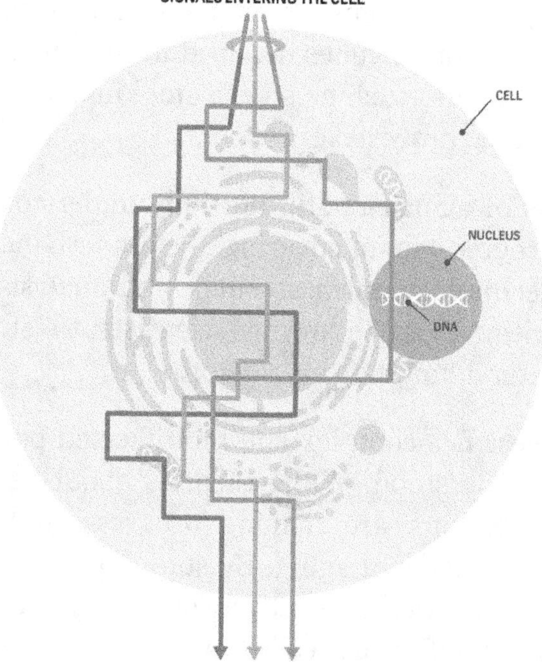

To answer questions about why and how we age, some scientists look for mechanisms or pathways in the body that lead to aging. Our cells constantly receive cues from both inside and outside the body, prompted by such things as injury, infection, stress, or even food. To react and adjust to these cues, cells send and receive signals through biological pathways. Some of the most common are involved in metabolism, the regulation of genes, and the transmission of signals. These pathways may also be important to aging.

Aging is a complex and multifaceted process that has been the subject of scientific study for centuries. In recent years, advances in technology and an increased understanding of the biology of aging have led to a surge in research into the topic. This booklet aims to provide an overview of some of the key areas of research into the biology of aging, highlighting how they fit together to help us better understand the aging process.

- Telomeres and Telomerase:

Telomeres are the protective caps at the end of chromosomes that shorten with each cell division. When telomeres become too short, the cell can no longer divide and will enter a state of senescence or undergo programmed cell death. Telomerase is an enzyme that can lengthen telomeres, and research into its function and regulation has been a major area of study in the biology of aging.

- Epigenetics:

Epigenetics is the study of changes in gene expression that do not involve changes to the DNA sequence itself. These changes can be influenced by a variety of factors, including aging, and can have a significant impact on cellular function. Research into epigenetics has revealed a complex landscape of gene regulation that is critical for understanding the aging process.

- Senescence:

Cellular senescence is a state in which a cell can no longer divide and is characterized by a distinctive secretory phenotype. Senescent cells accumulate with age and have been implicated in a variety of age-related diseases, including cancer, atherosclerosis, and neurodegenerative disorders. Research into senescence has focused on understanding its role in aging and developing therapeutic strategies to target senescent cells.

- Stem Cells and Regenerative Medicine:

Stem cells are undifferentiated cells that have the ability to give rise to multiple cell types. They play a critical role in tissue maintenance and regeneration, and their function declines with age. Research into stem cells and regenerative medicine has the potential to revolutionize our understanding of aging and could lead to the development of new therapies to promote tissue repair and regeneration.

- Mitochondria and Metabolism:

Mitochondria are the powerhouses of the cell, responsible for generating energy in the form of ATP. Mitochondrial function declines with age, leading to a decrease in cellular energy production and an increase in oxidative stress. Research into mitochondria and metabolism has focused on understanding the role of mitochondria in aging and developing strategies to improve mitochondrial function.

- Inflammation and Immunosenescence:

Inflammation is a natural response to tissue damage, but chronic inflammation can have negative consequences for health. Immunosenescence is the decline in immune function that occurs with age, making older adults more susceptible to infection and less able to fight off cancer. Research into inflammation and immunosenescence has focused on understanding their role in aging and developing strategies to promote healthy immune function.

- Hormones and Hormone Replacement Therapy:

Hormones play a critical role in regulating a variety of physiological processes, including growth and development, metabolism, and reproduction. Hormone levels decline with age, and this decline has been linked to a variety of age-related diseases. Research into hormones and hormone replacement therapy has focused on understanding the role of hormones in aging and developing strategies to replace hormones in order to promote healthy aging.

- Genomics and Personalized Medicine:

Genomics is the study of the complete set of genetic information for an organism. Personalized medicine is the tailoring of medical treatment to an individual's specific genetic profile. Research into genomics and personalized medicine has the potential to revolutionize our understanding of aging and could lead to the development of targeted therapies that are tailored to an individual's specific needs.

- Nutrition and Lifestyle:

Nutrition and lifestyle are critical factors in healthy aging. A healthy diet, regular exercise, and stress management can help to promote healthy aging and decrease the risk of age-related diseases.

What is Aging?

In the broadest sense, aging encompasses all the transformations that occur throughout life, from growth and development to maturity. The process of aging is multifaceted and can be viewed from various perspectives.

From a youthful perspective, aging is often associated with exciting changes, such as increased independence, later bedtimes, and curfews. However, as individuals reach middle age, the aging process can become more noticeable, with cosmetic changes like gray hair and wrinkles becoming more prominent.

Middle age is also a time when people begin to experience physical decline, even for those who are athletically fit. For instance, a study funded by the National Institute on Aging (NIA) found that marathon runners' record times increased with age, suggesting that aging can slow down even the most physically fit individuals.

While some physical decline may be a natural consequence of aging, the reasons behind these changes are of great interest to gerontologists. By studying the aging process, gerontologists aim to understand the underlying mechanisms that contribute to physical decline and identify ways to promote healthy aging.

Some of the key factors that contribute to physical decline with age include:

- Telomere shortening: Telomeres are the protective caps at the end of chromosomes that shorten with each cell division. As telomeres shorten, cells may reach a point where they can no longer divide, leading to cellular senescence.

- Epigenetic changes: Epigenetic changes refer to chemical modifications to DNA or histone proteins that can affect gene expression without altering the underlying DNA sequence. These changes can influence various biological processes and contribute to aging.

- Mitochondrial dysfunction: Mitochondria are the powerhouses of cells, responsible for generating energy in the form of ATP. Mitochondrial dysfunction can lead to a decline in energy production, which can contribute to physical decline.

- Inflammation: Chronic inflammation can contribute to aging by promoting oxidative stress and damaging cellular components.

- Hormonal changes: Hormonal changes can occur as we age, particularly the decline in hormones such as testosterone and estrogen, which can impact physical function and health.

- Lifestyle factors: Lifestyle factors such as smoking, excessive alcohol consumption, and lack of exercise can all contribute to physical decline.

Understanding these factors and their interplay is crucial for developing effective strategies to promote healthy aging and mitigate age-related diseases. By studying the aging process and identifying the underlying mechanisms that contribute to physical decline, gerontologists can help to improve the health and well-being of older adults.

Gerontologists are interested in understanding the underlying mechanisms that contribute to the aging process and the factors that distinguish normal aging from disease. They also seek to identify why older adults are more susceptible to disease and disability, and why these health threats have a greater impact on older bodies.

To answer these questions, the Baltimore Longitudinal Study of Aging (BLSA) has been conducting research since 195 - The BLSA is a longitudinal study, which means that it repeatedly evaluates the same individuals over time, rather than comparing a group of young people to a group of older people, as in a cross-sectional study.

One of the key findings from the BLSA is that people who have no evidence of ear problems or noise-induced hearing loss still experience some hearing loss as they age. This is considered a normal part of the aging process.

Another important discovery made by BLSA scientists is that even healthy individuals who maintain good brain function throughout their lives lose a significant amount of brain volume during normal aging. This loss of brain volume can be detected using brain scans, and it is thought to be related to cognitive changes that occur with age.

These findings highlight the importance of understanding the aging process and the factors that contribute to it. By studying normal aging and comparing it to disease, gerontologists can identify potential targets for interventions that may help to promote healthy aging and prevent age-related diseases. Additionally, the use of longitudinal studies like the BLSA provides valuable insights into the changes that occur over time, allowing researchers to better understand the aging process and develop effective strategies to promote healthy aging.

It's interesting to note that some changes that were previously thought to be a normal part of the aging process, such as changes in personality, may actually be early signs of disease or dementia. This highlights the importance of monitoring and understanding the aging process, in order to identify potential health issues before they become more serious.

One way to do this is by studying the biology of aging, and looking for mechanisms or pathways in the body that may contribute to the aging process. For example, scientists have found that cells in the body are constantly receiving cues from both inside and outside the body, and that these cues can affect the aging process. By understanding these cues and the pathways they trigger, scientists may be able to develop strategies to promote healthy aging and prevent age-related diseases.

One possible pathway that has been identified is the regulation of genes, which can be influenced by factors such as stress, injury, and infection. Another pathway is the transmission of signals between cells, which can be affected by factors such as hormonal changes and inflammation. By studying these pathways and how they interact with each other, scientists may be able to identify potential targets for interventions that can promote healthy aging.

It's also worth noting that some scientists believe that aging may not be a purely biological process, but may also be influenced by social and environmental factors. For example, studies have shown that people who are socially active and engaged in their communities tend to live longer and healthier lives than those who are isolated and inactive. This suggests that factors such as social support and engagement may also play a role in the aging process.

It's clear that the aging process is complex and multifaceted, and that there is still much to be learned about it. By continuing to study and understand the biology of aging, scientists may be able to develop new strategies to promote healthy aging and improve the quality of life for older adults.

The rate and progression of cellular aging can vary greatly from person to person, but it is a natural process that affects the cells of every major organ in the body over time. Some changes can start early and have a significant impact on health and function, while others may have less of an impact.

For example, around the age of 20, lung tissue starts to lose elasticity, and the muscles of the rib cage slowly begin to shrink, which can decrease the maximum amount of air that can be inhaled. This can make it more difficult to breathe and may lead to respiratory problems later in life.

In the gut, production of digestive enzymes diminishes with age, which can affect the ability to absorb food properly and maintain a nutritional balance. This can lead to digestive problems and nutrient deficiencies if left unchecked.

Blood vessels in the heart accumulate fatty deposits and lose flexibility over time, resulting in atherosclerosis, also known as "hardening of the arteries." This can increase the risk of heart disease, heart attack, and stroke.

In women, vaginal fluid production decreases with age, and sexual tissues can atrophy, leading to discomfort during sex and increased risk of infections. In men, aging can decrease sperm production, and the prostate can become enlarged, leading to urinary problems.

It's important to note that while these changes can have a significant impact on health and quality of life, there are steps that can be taken to slow down or even prevent some of these changes. A healthy diet, regular exercise, and regular check-ups with a healthcare provider can help to maintain health and well-being as we age.

Scientists are making significant progress in understanding the age-related differences that occur in our bodies, thanks to studies like the Baltimore Longitudinal Study of Aging (BLSA). The BLSA is a long-term study that has been tracking the aging process in a group of individuals over several decades. By analyzing data from this study and others like it, scientists are able to identify patterns and trends in the changes that occur in our bodies as we age.

One of the key ways that scientists are able to study aging is by comparing young and old individuals. By comparing the cells and tissues of young and old individuals, scientists can identify differences in the way that genes are expressed, how cells function, and how tissues are structured. For example, scientists have found that the cells of older individuals tend to have more damage to their DNA and membranes, and that their mitochondria (the powerhouses of the cell) are less efficient at producing energy.

However, while these studies can tell us a lot about how aging affects our bodies, they don't necessarily tell us why aging happens in the first place. There are still many unanswered questions about what triggers aging in our cells and tissues, and what biological processes are involved.

To get a deeper understanding of aging, scientists are looking at the cellular and molecular levels. They are studying the changes that occur in our cells and in laboratory animals to identify the underlying causes of aging. For example, scientists have found that certain genes are involved in the aging process, and that the expression of these genes can be influenced by factors such as diet, exercise, and stress.

By studying aging at the cellular and molecular levels, scientists hope to develop new ways to slow down or even reverse the aging process. For example, they may be able to develop drugs or therapies that target specific genes or biological processes involved in aging. They may also be able to develop strategies for maintaining health and preventing age-related diseases, such as cancer, heart disease, and Alzheimer's.

Ultimately, the goal of this research is to help people live longer, healthier lives. By understanding the underlying causes of aging, scientists hope to be able to develop interventions that can help us maintain our health and well-being as we age. This could have a significant impact on public health, as it could help to reduce the burden of age-related diseases on our healthcare system and improve the quality of life for millions of people.

Living long and well: Can we do both? Are they the same?

The concepts of lifespan and health span are closely related but distinct. Lifespan refers to the length of time that an individual lives, whereas health span refers to the period of life during which an individual is healthy and free from disease. While it is possible for an individual to have a long lifespan, their health span may not necessarily be equally long. For example, an individual may live to be 100 years old, but spend the last 20 years of their life in poor health, with chronic illnesses or disabilities.

Researchers are working to understand the aging process and identify ways to extend health span, rather than just lifespan. They are studying various factors that contribute to aging, such as genetics, lifestyle choices, and environmental factors, and looking for ways to slow down or reverse the aging process. Some of the strategies being explored include:

 - Caloric restriction: Reducing daily caloric intake has been shown to increase lifespan and health span in some animal studies.

 - Exercise: Regular exercise has been linked to improved health outcomes and a lower risk of chronic diseases.

 - Diet: Eating a healthy, balanced diet that is rich in nutrients can help support health and well-being.

- Stress reduction: Chronic stress can have negative effects on health, so finding ways to manage stress, such as through meditation or yoga, can be beneficial.

- Sleep: Getting enough sleep is essential for health and well-being, and can help to reduce the risk of chronic diseases.

- Social connections: Maintaining social connections and building strong relationships can help to support mental and emotional health.

- Medications and therapies: Researchers are also exploring various medications and therapies that can help to extend health span, such as drugs that target specific aging mechanisms and stem cell therapies.

While it is unlikely that a single "anti-aging" elixir will be discovered, researchers believe that their work will reveal ways to improve a person's ability to live a longer, healthier life. By understanding the aging process and identifying strategies to extend health span, individuals can take steps to live a healthier, more fulfilling life, and improve their quality of life.

FACTORS CONTRIBUTING TO LIFESPAN

GENES

ENVIRONMENT

BEHAVIORAL TRAITS

Lifespan refers to the length of time that an organism lives, and it can vary depending on various factors such as species, genetics, lifestyle, and environmental conditions. Maximal lifespan, on the other hand, refers to the maximum number of years of life observed in a specific population, and it can be used to compare the lifespan of different species or populations.

It's worth noting that the maximum recorded lifespan for humans is not necessarily the average lifespan, which is the average number of years that a person is expected to live based on their age, gender, and other factors. The average lifespan for humans varies depending on factors such as where they live, their lifestyle, and their access to healthcare.

In any case, it's fascinating to learn about the different ways that species age and the factors that influence their lifespan. From telomeres to genetics to lifestyle choices, there's still much to be discovered and explored in the field of aging and longevity!

Lifespan is a widely used measurement in aging research because it provides a clear and objective way to assess the effect of various factors on aging. By studying the lifespan of organisms, scientists can identify factors that contribute to aging and understand how they influence the aging process.

One common approach to studying aging is to manipulate a specific factor, such as a gene or environmental condition, and measure its effect on lifespan. This can help researchers determine whether the factor is important for aging and can provide insights into the underlying mechanisms of aging.

For example, researchers may modify the activity of a gene suspected to be involved in aging, such as by deleting or duplicating the gene, and measure the effect on lifespan. If the modified gene activity leads to a longer or shorter lifespan, it suggests that the gene plays a role in aging.

Studying lifespan is a powerful tool for understanding the aging process and identifying potential therapeutic targets for age-related diseases.

It's exciting to hear about the progress being made in the field of aging research. The fact that external factors, such as dietary supplements, hormones, and anti-inflammatory drugs, may influence lifespan is a promising area of study. The findings from the Interventions Testing Program (ITP) studies on mice are particularly intriguing, especially the discovery that aspirin and masoprocol may have a positive impact on lifespan.

However, it's important to note that more research is needed to fully understand the effects of these compounds on humans. While some of the compounds tested in the ITP already have clinical use for humans, it's crucial to use them only as prescribed and not for lifespan extension at this time.

I'm curious, what are some of the potential risks associated with using these compounds for lifespan extension in humans? And what are some of the ethical considerations that need to be taken into account when it comes to extending human lifespan?

The ability to withstand disease is a crucial factor in determining lifespan. Research on exceptionally long-lived individuals has shown that health problems tend to be delayed, appearing closer to the end of life, rather than occurring at the same age in all people. This suggests that some individuals may be able to prolong their healthspan, or the period of life spent in good health, beyond the average lifespan.

A Danish longitudinal study of 92- to 100-year-olds found that health problems seem to be delayed in this population, with the average centenarian appearing to be in better health than the average 80-year-old. However, it's important to note that living to 100 does not mean never having any health issues.

In the New England Centenarian Study, researchers have developed three categories for their long-lived participants: "survivors," "delayers," and "escapers." Survivors are individuals who have survived a life-threatening disease, while delayers are those who have delayed a serious health problem until much later in life. Escapers, on the other hand, are individuals who have escaped any serious health events throughout their lives.

These categories suggest that there may be different strategies for achieving a long and healthy life. Some individuals may be able to survive serious health problems through access to good healthcare, while others may be able to delay the onset of health issues through healthy lifestyle choices. Still, others may be able to escape serious health problems altogether due to genetic or environmental factors.

Understanding the factors that contribute to a long and healthy life is crucial for improving public health and increasing lifespan. By studying exceptionally long-lived individuals, researchers can identify patterns and strategies that may help to promote healthy aging and extend lifespan.

This text suggests that resveratrol, a compound found in wine, may have positive effects on the health of mice, even if it does not increase their lifespan. The study mentioned found that resveratrol-treated mice had better bone health, heart function, strength, vision, coordination, and cholesterol levels than the control group. However, it is important to note that this study was conducted on mice, and it is unclear whether the same effects would occur in humans. Additionally, the study found that resveratrol did not increase lifespan in mice on a normal diet, but it did increase lifespan in mice on a high-fat diet. This suggests that the effects of resveratrol on lifespan may depend

UNCOVERING FAMILY SECRETS TO A LONG LIFE

Most of what we know about factors that can contribute to a long lifespan and health span is based on research in animal models. However, NIA-funded research like the Long Life Family Study is taking what we've learned in animals and seeing if it applies to human aging. This study is collecting data from families with at least two siblings who have lived to a very old age in relatively good health. Along with asking questions about their family and health history, the researchers conduct physical assessments and health screenings and collect a small blood sample for genetic tests. What researchers learn about common characteristics shared by these families could one day be used to guide lifestyle advice and medical treatments.

on the individual's diet and health status. Further research is needed to fully understand the potential health benefits and risks of resveratrol in humans.

Imagine a society where people live to be 100, but with a significant decline in physical health. This scenario highlights the importance of health span research, which focuses on understanding how to delay or prevent disease and disability, allowing individuals to live healthier for longer. While lifespan research is still valuable, health span research has the potential to reveal strategies for maintaining good health and preventing age-related diseases, ultimately improving the quality of life for individuals as they age. By shifting the focus to health span, researchers may uncover ways to extend the period of healthy life, enabling people to live well into their 80s, 90s, and beyond, without the burden of debilitating illnesses.

Is what's good for mice good for men?

Animal models are an essential tool in research on aging and age-related diseases. Organisms like fruit flies, roundworms, mice, and nonhuman primates share many biological mechanisms and genes with humans, making them useful for studying human physiology and aging. These animals can be used to model human aging and age-

ANIMAL MODELS

Here are some animals commonly studied in aging research.

ROUNDWORM
C. elegans

FRUIT FLY
Drosophila melanogaster

MOUSE
Mus musculous

RHESUS MONKEY
Macaca mulatta

related diseases, allowing scientists to explore the effects of genetic modifications, dietary interventions, and other variables on health and longevity.

One advantage of using animal models is that they can be studied in tightly controlled environments, which allows researchers to isolate the variable they want to investigate. This is particularly useful when studying the effects of genetic modifications or dietary interventions, as it allows scientists to rule out the effects of other factors that could influence the results.

Moreover, animal models enable scientists to conduct experiments that would not be possible in humans. For example, researchers can modify a gene in a mouse or a worm to study its effects on health and longevity, which would not be ethical or practical in humans.

However, it's important to note that animal models have limitations. For instance, the biology of aging in animals can differ from that in humans, and results from animal studies may not always translate to humans. Additionally, animal models may not capture the full complexity of human aging, which can be influenced by a wide range of factors, including lifestyle, environment, and genetics.

Therefore, while animal models are a valuable tool for understanding aging and age-related diseases, they should be used in conjunction with other research methods, such as observational studies and clinical trials in humans. By combining data from multiple approaches, scientists can gain a more comprehensive understanding of the aging process and develop effective interventions to promote healthy aging.

Different types of studies use different animal models, each with its own advantages and limitations. Animal models with a short lifespan, such as fruit flies and roundworms, offer several benefits for studying aging and longevity.

One advantage of using short-lived animal models is that they require less time and resources to study from birth to death. This allows researchers to measure the effects of interventions in a relatively short period, which can be beneficial for studying the aging process. For example, scientists might favor using fruit flies when studying a possible genetic target for an intervention to increase longevity because their average lifespan is only 30 days. This allows researchers to measure the effects in about a month.

A DIFFERENT APPROACH: COMPARATIVE BIOLOGY

MOUSE
Lifespan: 4 years

NAKED MOLE RAT
Lifespan: 17 to 28 years

One approach to aging biology research is called "comparative biology." It involves comparing two or more similar species that have very different lifespans—one lives much longer than the other—to understand how the longer-lived species has, as one NIA-funded researcher puts it, "exceptional resistance to basic aging processes." Comparative biology studies generally focus on species that live at least twice as long as their close relatives.

A few possible theories explain what may be taking place among these longer-lived animals:

- They experience a slower rate of age-related decline.
- They can survive even when their organs and/or systems break down and have minimal function.
- They are better able to tolerate cellular damage or diseases.

The naked mole rat, a mouse-size rodent that lives underground, has been widely used in comparative research. It lives approximately 17 years in the wild and more than 28 years in captivity. Its relative, the mouse, lives a maximum of 4 years. What accounts for this startling difference? Naked mole rats have lower metabolic rates and body temperature, meaning that they require less energy to survive. They have low concentrations of blood glucose (blood sugar), insulin, and thyroid hormone, so they are less susceptible to certain diseases. Naked mole rats are better able to withstand some types of biological stress and, at this point, there has never been a case of cancer reported in these animals. All these factors and likely others yet to be determined contribute to their healthier and longer life.

Another advantage of short-lived animal models is that they can be used to study genes that might affect longevity in a more efficient manner. The roundworm's 2- to 3-week lifespan makes it an ideal model for identifying and studying genes that might affect longevity. In a landmark study, NIA-funded researchers found that reducing the activity of a set of genes, called daf, increased roundworm lifespan by three- or even fourfold. This research would not have been as feasible if conducted using an animal model with an average lifespan of 10 or 20 years.

However, it is important to note that short-lived animal models also have some limitations. For example, their short lifespan might not accurately reflect the aging process in longer-lived organisms, such as humans. Additionally, short-lived animal models might not be suitable for studying certain age-related diseases that take years or decades to develop.

Different types of studies use different animal models, each with its own advantages and limitations. Short-lived animal models, such as fruit flies and roundworms, offer several benefits for studying aging and longevity, including reduced time and resources required for study, ability to study genes that might affect longevity in a more efficient manner, and ability to measure the effects of interventions in a relatively short period. However, it is important to consider the limitations of short-lived animal models when interpreting the results of studies that use them.

When scientists are studying aging and trying to find ways to extend lifespan, they often start by studying simple organisms like worms or flies. These organisms are easy to work with and have a relatively short lifespan, which makes it easier to study the effects of aging and potential interventions.

Once scientists have identified a possible intervention that seems to work in these simple organisms, they then try to apply it to more complex organisms, like mice. Mice are a good model for human aging because they have a similar number of genes to humans and their genes serve similar functions. However, the activity of these genes is different in mice and humans, which can affect how well an intervention works.

For example, a genetic intervention that increases a worm's lifespan by fourfold might only have a small effect on a mouse's lifespan. This is because the genes that control aging in worms and mice are different, and the intervention might not be able to target the right genes in mice.

Similarly, an intervention that works well in mice might not work at all in humans. This is because humans have a much more complex genome than mice, with many more genes that are involved in aging. Additionally, the activity of these genes is highly regulated and can be influenced by a wide range of factors, including lifestyle, environment, and genetics.

Therefore, it's important for scientists to carefully study the effects of aging interventions in a variety of organisms, including simple models like worms and flies, as well as more complex models like mice and nonhuman primates. By doing so, they can gain a better understanding of how these interventions work and whether they are likely to be effective in humans.

Studies in animal models closer to humans, such as monkeys or other nonhuman primates, play a crucial role in understanding how basic discoveries made in earlier animal models might apply to humans. These studies are essential for pre-clinical studies, which serve as an intermediary step between research in animal models like mice and clinical studies in humans.

One key area where studies in nonhuman primates have been particularly informative is in understanding how normal age-related changes in the heart influence the risk of heart disease. For example, researchers have used nonhuman primates to study the effects of aging on the heart and how these changes can increase the risk of heart disease. They have also used these models to test interventions aimed at lowering the risk of heart disease, such as drugs that decrease blood vessel stiffness.

The importance of studying aging in nonhuman primates lies in the fact that their aging processes are similar to those in humans. As a result, findings from these studies can provide valuable insights into the underlying mechanisms of aging and age-related diseases in humans. Additionally, nonhuman primates are closer to humans in terms of their genetic, physiological, and behavioral characteristics, making them a more reliable model for studying the effects of aging and testing interventions.

For instance, studies in rhesus macaques have shown that age-related changes in the heart, such as increased stiffness of blood vessels, can lead to an increased risk of heart disease. These changes can be reversed or prevented with drugs that target specific molecular pathways, such as the renin-angiotensin-aldosterone system (RAAS). These findings have important implications for the development of therapies aimed at reducing the risk of heart disease in humans.

Furthermore, studies in nonhuman primates have also been used to test the safety and efficacy of drugs aimed at treating age-related diseases, such as Alzheimer's disease, Parkinson's disease, and cancer. These studies have shown that certain drugs can improve cognitive function, reduce inflammation, and slow the progression of disease in nonhuman primates, providing hope for their potential use in humans.

Studies in animal models closer to humans, such as nonhuman primates, are essential for understanding how basic discoveries made in earlier animal models might apply to humans. These studies provide valuable insights into the underlying mechanisms of aging and age-related diseases and are a crucial step in the development of therapies aimed at improving human health.

Studies in animal models closer to humans, such as monkeys or other nonhuman primates, play a crucial role in understanding how basic discoveries made in earlier animal models might apply to humans. These studies are essential for pre-clinical studies, which serve as an intermediary step between research in animal models like mice and clinical studies in humans.

One key area where studies in nonhuman primates have been particularly informative is in understanding how normal age-related changes in the heart influence the risk of heart disease. For example, researchers have used nonhuman primates to study the effects of aging on the heart and how these changes can increase the risk of heart disease. They have also used these models to test interventions aimed at lowering the risk of heart disease, such as drugs that decrease blood vessel stiffness.

The importance of studying aging in nonhuman primates lies in the fact that their aging processes are similar to those in humans. As a result, findings from these studies can provide valuable insights into the underlying mechanisms of aging and age-related diseases in humans. Additionally, nonhuman primates are closer to humans in terms of their genetic, physiological, and behavioral characteristics, making them a more reliable model for studying the effects of aging and testing interventions.

For instance, studies in rhesus macaques have shown that age-related changes in the heart, such as increased stiffness of blood vessels, can lead to an increased risk of heart disease. These changes can be reversed or prevented with drugs that target specific molecular pathways, such as the renin-angiotensin-aldosterone system (RAAS). These findings have important implications for the development of therapies aimed at reducing the risk of heart disease in humans.

Furthermore, studies in nonhuman primates have also been used to test the safety and efficacy of drugs aimed at treating age-related diseases, such as Alzheimer's disease, Parkinson's disease, and cancer. These studies have shown that certain drugs can improve cognitive function, reduce inflammation, and slow the progression of disease in nonhuman primates, providing hope for their potential use in humans.

Studies in animal models closer to humans, such as nonhuman primates, are essential for understanding how basic discoveries made in earlier animal models might apply to humans. These studies provide valuable insights into the underlying mechanisms of aging and age-related diseases and are a crucial step in the development of therapies aimed at improving human health.

When scientists study aging in animals, they're looking for ways to slow down or reverse the aging process. They might test different interventions, such as drugs or dietary changes, to see if they have any effect on aging. If an intervention works in an animal model, it can be promising for future human studies.

However, it's important to remember that animals are not humans. While animal models can provide valuable insights into aging, they don't always translate to humans. There are many reasons for this:

Animals have different biology than humans. For example, mice have a much shorter lifespan than humans, so their aging process is different.

Animals may not have the same diseases or health issues as humans. For example, mice may not develop the same type of cancer as humans.

Animals may respond differently to interventions than humans. For example, a drug that works well in mice may not work as well in humans.

Animals may have different lifestyles than humans. For example, mice may not eat the same foods as humans, and they may not have the same level of physical activity.

Because of these differences, it's important to do human studies to test interventions that have shown promise in animal models. Human studies can help scientists understand how an intervention works in humans, what the side effects are, and whether it's safe and effective.

In addition, human studies can also help scientists identify potential risks or complications that may not have been seen in animal models. For example, a drug that works well in mice may cause unexpected side effects in humans, such as liver damage or allergic reactions.

So while animal studies are an important step in the scientific process, they're not the final step. Scientists need to do human studies to confirm that an intervention is safe and effective before it can be widely adopted.

GENETICS

Is aging in our genes?

Yes, aging is influenced by our genes to some extent. Research has shown that certain genetic variations can affect a person's lifespan and susceptibility to age-related diseases. For example, some families have a higher risk of developing certain age-related diseases, such as Alzheimer's or Parkinson's, due to genetic mutations. Additionally, some people may have a genetic predisposition to age-related changes in physical function or cognitive decline.

However, it's important to note that genetics is not the only factor that contributes to aging. Lifestyle choices, environmental factors, and random chance also play a role in how we age. While some people may be born with genetic mutations that predispose them to age-related diseases, others may have a genetic makeup that protects them from such diseases.

The study of genetics and aging is an active area of research, with scientists working to identify the specific genetic factors that contribute to aging and age-related diseases. By understanding the genetic basis of aging, researchers hope to develop new treatments and interventions that can help people live healthier and longer lives.

While genetics does play a role in aging, it's not the only factor that contributes to how we age. A combination of genetic, lifestyle, and environmental factors all influence the aging process, and researchers are working to understand the complex interplay between these factors.

Identifying the genes associated with a particular trait can be a challenging task. It requires a thorough understanding of the trait, including the various factors and pathways that contribute to it. This knowledge is essential for identifying the specific genes that are likely to be involved in the trait.

Once a gene has been identified as a potential contributor to the trait, scientists must then determine the nature of its relationship with the trait. This can be a complex process, as the relationship between a gene and a trait can be direct, indirect, or even non-existent.

To establish a direct relationship between a gene and a trait, scientists typically use genetic association studies. These studies involve comparing the genetic variations in individuals with the trait to those without the trait. If a specific genetic variation is found to be more common in individuals with the trait, it suggests that the gene associated with that variation may be responsible for the trait.

However, it's not always possible to identify a direct relationship between a gene and a trait. In some cases, the relationship may be indirect, meaning that the gene influences the trait through a complex interplay of other genetic and environmental factors. In other cases, the relationship may be non-existent, meaning that the gene does not play a role in the trait.

To further complicate matters, genetic variations can interact with environmental factors in complex ways, making it difficult to tease apart the relative contributions of genetics and environment to a particular trait.

Despite these challenges, researchers have made significant progress in identifying the genes associated with a wide range of traits, including common diseases such as heart disease, diabetes, and mental health disorders. Advances in genomics and other technologies have enabled scientists to analyze large amounts of genetic data and identify patterns and relationships that were previously unknown.

However, much work remains to be done. Identifying the genes associated with a particular trait is just the first step in understanding the underlying biology and developing effective treatments. Further research is needed to fully understand the complex interplay of genetic and environmental factors that contribute to a particular trait and to develop targeted interventions that can effectively modify the trait.

Identifying longevity genes is a complex task, indeed. Unlike traits such as height or hair color, which are influenced by a limited number of genes, longevity is a complex trait that is influenced by numerous genetic and environmental factors.

Scientists have made progress in identifying genes that affect longevity in short-lived animal models, such as worms and flies. These models have a shorter lifespan than humans, which allows researchers to study multiple generations and identify genes that influence lifespan.

However, it's important to note that not all genes that affect longevity in these models are necessarily beneficial. Some genes may actually limit lifespan, and mutating or eliminating them may increase lifespan. This suggests that the normal function of these genes may be detrimental to longevity.

Findings in animal models can provide valuable insights into the genes that may influence longevity in humans. However, it's important to recognize that the genetic and environmental factors that contribute to longevity in humans are likely to be more complex and multifaceted than those in animal models.

Therefore, while animal models can provide useful information, it's essential to conduct further research in humans to identify the genes that truly influence longevity. This may involve studying large populations of people over an extended period, as well as using advanced genomic techniques to analyze genetic data.

Ultimately, identifying the genes that influence longevity in humans will require a multidisciplinary approach that integrates insights from genetics, genomics, and other fields. By understanding the genetic and environmental factors that contribute to longevity, scientists may be able to develop targeted interventions that promote healthy aging and extend human lifespan.

How can we find aging genes in humans?

There are several ways to identify aging genes in humans:

- Genome-wide association studies (GWAS): This approach involves scanning the entire genome to identify genetic variants that are associated with a particular trait or disease. GWAS can be used to identify genes that are associated with longevity or aging-related traits, such as resistance to age-related diseases.

- Whole-genome sequencing: This approach involves sequencing the entire genome of a large number of individuals to identify genetic variants that are associated with a particular trait or disease. Whole-genome sequencing can be used to identify rare genetic variants that may be associated with aging-related traits.

- Candidate gene studies: This approach involves studying specific genes that are known to be involved in aging-related processes, such as cellular senescence, telomere shortening, or oxidative stress. Candidate gene studies can be used to identify genetic variants that are associated with aging-related traits.

- Genetic linkage analysis: This approach involves studying the genetic patterns of family members to identify genetic variants that are associated with a particular trait or disease. Genetic linkage analysis can be used to identify genes that are associated with aging-related traits.

- Expression quantitative trait loci (eQTL) analysis: This approach involves studying the genetic variants that are associated with the expression levels of specific genes. eQTL analysis can be used to identify genes that are associated with aging-related traits.

- machine learning algorithms: These algorithms can be used to identify patterns in large datasets, including genomic data. Machine learning algorithms can be used to identify genes that are associated with aging-related traits.

- Systems biology approaches: These approaches involve studying the interactions between genes, proteins, and other molecules in a systematic way. Systems biology approaches can be used to identify genes that are associated with aging-related traits.

- Computer models and simulations: These can be used to predict the behavior of genes and their interactions with other molecules, and can be used to identify potential aging-related genes.

- Comparative genomics: This approach involves comparing the genomes of different species to identify genes that are conserved across different species. Comparative genomics can be used to identify genes that are associated with aging-related traits.

- Epigenomics: This approach involves studying the epigenetic modifications, such as DNA methylation and histone modifications, that can affect gene expression. Epigenomics can be used to identify genes that are associated with aging-related traits.

It is important to note that no single approach is guaranteed to identify all aging genes, and that a combination of approaches is likely to be more effective. Additionally, identifying genes associated with aging is only the first step, and understanding the functional mechanisms by which these genes influence aging will require further study.

The candidate gene approach is a method used to identify genes that may be associated with longevity in humans. This approach involves identifying genes in humans that have similar functions to genes that have already been linked to aging in animal models. These genes are called "homologs" or "orthologs" to animal genes.

The candidate gene approach is based on the idea that genes that are involved in aging-related processes in animal models may also be involved in aging-related processes in humans. By identifying genes that are similar in function to those in animal models, scientists can explore whether these genes are also associated with longevity in humans.

To do this, scientists first identify genes that are involved in aging-related processes in animal models. For example, they may look at genes that are involved in the insulin/IGF-1 pathway, which has been linked to aging in animal models. Then, they search for the comparable genes in the insulin/IGF-1 pathway of humans.

Once they have identified the human genes that are similar to the animal genes, they determine whether these genes are associated with longevity in humans. They do this by looking for variants of the genes that are prevalent among people who live healthy, long lives but not for people who have an average health span and lifespan.

The candidate gene approach is a way for scientists to identify potential genes associated with longevity in humans by looking at genes that have already been linked to aging-related processes in animal models.

In a project funded by the National Institute on Aging (NIA), researchers investigated the relationship between genes associated with the insulin/IGF-1 pathway and longevity in humans. The study focused on 30 genes that are known to play a role in this pathway and analyzed their variants in a group of women over 92 years old compared to a group of women under 80 years old.

The researchers found that certain variants of the FOXO3a gene were more common among the long-lived individuals, suggesting a potential association with longer lifespan. This finding supports the idea that the insulin/IGF-1 pathway, which has been linked to aging in animal models, also plays a role in human aging.

The discovery of genes that may be important for healthy aging provides valuable insights for the development of therapies to support aging and age-related diseases. Understanding the genetic factors that contribute to longevity can help researchers identify potential targets for interventions that may promote healthy aging and improve quality of life for older adults.

This study highlights the potential of genetic research to uncover new ways to support healthy aging and improve outcomes for older adults. By continuing to explore the genetic factors that contribute to longevity, researchers may be able to develop new therapeutic approaches that can help promote healthy aging and improve quality of life for older adults.

The genome-wide association study (GWAS) is a powerful approach in identifying genes associated with diseases and conditions related to aging. GWAS involves scanning the entire genome to identify variants that occur more frequently in a group with a particular health issue or trait.

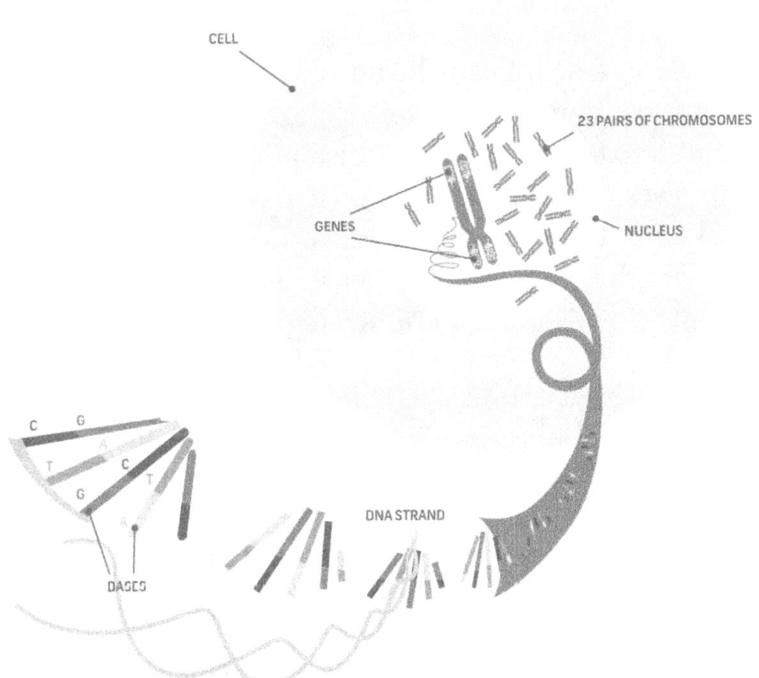

The human genetic blueprint, or genome, consists of approximately 25,000 genes made up of approximately 3 billion letters (base pairs) of DNA. Base pair sequences: guanine (G) pairs with cytosine (C); adenine (A) pairs with thymine (T).

In a GWAS study funded by the National Institutes of Health (NIH), researchers identified genes possibly associated with high and low blood fat levels, cholesterol, and risk for coronary artery disease. The study analyzed data collected from a small, genetically similar population living in Sardinia, Italy, as well as from two other international studies.

The findings revealed more than 25 genetic variants in 18 genes connected to cholesterol and lipid levels. Notably, seven of these genes were not previously known to be associated with cholesterol and lipid levels, suggesting that there may be other pathways linked to risk for coronary artery disease.

Heart disease is a significant health concern for older adults, and finding ways to eliminate or lower risk for this condition could have a profound impact on reducing disability and death from age-related diseases. The identification of new genes associated with cholesterol and lipid levels provides valuable insights into the genetic factors that contribute to heart disease and may lead to the development of novel therapeutic strategies for reducing risk.

Moreover, the use of GWAS can help identify individuals who are at high risk for heart disease, allowing for early intervention and prevention strategies. This approach can also help researchers understand the underlying biology of heart disease, which may lead to the development of new treatments and therapies.

GWAS is a productive approach in finding genes involved in diseases and conditions associated with aging, such as heart disease. The identification of new genetic variants associated with cholesterol and lipid levels provides valuable insights into the genetic factors that contribute to heart disease and may lead to the development of novel therapeutic strategies for reducing risk. The use of GWAS can also help identify individuals who are at high risk for heart disease, allowing for early intervention and prevention strategies.

GWAS (Genome-Wide Association Study) is a powerful tool for identifying genes associated with aging and longevity. Unlike the candidate gene approach, which requires prior knowledge of the function of the gene and its potential relationship with longevity, GWAS does not require any prior knowledge. This makes GWAS a valuable tool for discovering novel genes involved in aging that may not have been previously considered.

GWAS works by scanning the entire genome to identify single nucleotide polymorphisms (SNPs) that are associated with a particular trait or condition, in this case, aging and longevity. The study looks for patterns of genetic variation that are more common in individuals who live longer or healthier lives, compared to those who do not. By analyzing large amounts of data from diverse populations, GWAS can help identify genes that may play a role in aging and longevity.

One of the main advantages of GWAS is that it can uncover genes involved in cellular processes and pathways that were not previously thought to play roles in aging. For example, a GWAS study published in the journal Nature in 2017 identified a genetic variant associated with longevity in the species C. elegans, a nematode worm. The study found that the variant was linked to a gene involved in the regulation of a cellular process called autophagy, which is the process by which cells recycle damaged or dysfunctional components. This discovery suggests that autophagy may play a role in aging and longevity, and highlights the potential of GWAS to uncover novel genes involved in aging.

Because no single approach can precisely identify each and every gene involved in aging, scientists often use multiple methods, including a combination of the GWAS and candidate gene approaches, to identify genes involved in aging. By combining different approaches, researchers can increase the power and accuracy of their findings, and gain a more comprehensive understanding of the genetic factors that contribute to aging and longevity.

GWAS is a valuable tool for identifying genes directly associated with aging and longevity, and has the potential to uncover novel genes involved in cellular processes and pathways that were not previously thought to play roles in aging. By combining GWAS with other approaches, scientists can gain a deeper understanding of the genetic factors that contribute to aging and develop new strategies for promoting healthy aging and extending lifespan.

GWAS is a powerful tool for identifying genes associated with aging and longevity, and has the potential to uncover novel genes involved in cellular processes and pathways that were not previously thought to play roles in aging. By combining GWAS with other approaches, scientists can gain a deeper understanding of the genetic factors that contribute to aging and develop new strategies for promoting healthy aging and extending lifespan.

As researchers delve deeper into the genetics of aging, they are discovering that the process is much more complex than previously

---

PATHWAYS OF LONGEVITY GENES

Most longevity genes identified thus far influence one of three pathways in a cell: insulin/IGF-1, sirtuins, or mTOR.

In the 1980s, scientists discovered the first gene shown to limit lifespan in roundworms, which they named *age-1*. Further investigation revealed that the effects of *age-1* are involved with the insulin/IGF-1 pathway. When scientists "silenced" the *age-1* gene's activity, the insulin/IGF-1 pathway's activity also decreased and the worms lived longer. Since then, many other genes associated with the insulin/IGF-1 pathway have been found to affect the lifespan of fruit flies and mice, strengthening the hypothesis that the insulin/IGF-1 pathway plays an important role in the aging process. More research is needed to determine if inhibiting this pathway could increase longevity in humans or create insulin-related health problems like diabetes. A recent report suggests that people with a mutation related to the insulin/IGF-1 pathway may have less risk of developing diabetes and cancer.

There is also a great deal of interest in the sirtuin pathway. Sirtuin genes are present in all species and regulate metabolism in the cell. They are crucial for cell activity and cell life. In the 1990s, scientists at the Massachusetts Institute of Technology found that inserting an extra copy of a sirtuin equivalent, called *Sir2*, increased the lifespan of yeast. Extension of lifespan has been replicated in other organisms, including flies and worms. However, studies in mice have yielded conflicting results.

The mTOR pathway—an abbreviation of "mammalian target of rapamycin"—plays a role in aging of yeast, worms, flies, and mice. This pathway controls the cell's rate of protein synthesis, which is important for proper cell function. Researchers have found that inhibiting the pathway in mice genetically or pharmacologically (using rapamycin) leads to increased longevity and improved health span.

thought. It is becoming increasingly clear that longevity is influenced by a multitude of genetic factors, each playing a small role in the aging process.

One way that genes may influence longevity is by affecting an individual's ability to survive diseases. Some people may have genes that make them more resistant to certain diseases, allowing them to live longer. Others may have genes that predispose them to certain diseases, which could shorten their lifespan.

Another way that genes may influence longevity is by affecting the rate at which we age. Some genes may accelerate the aging process, while others may slow it down. For example, researchers have identified genes that influence the production of telomerase, an enzyme that helps to protect the ends of chromosomes from damage. People with longer telomeres tend to live longer, healthier lives.

However, it is unlikely that scientists will identify a single "Eureka!" moment when one gene is discovered as the principal factor affecting health and lifespan. Instead, it is more likely that we will identify several combinations of many genes that affect aging, each to a small degree.

For example, a person may have a combination of genes that make them resistant to certain diseases, as well as genes that slow down the aging process. Another person may have a different combination of genes that accelerate the aging process, but also have genes that help them survive diseases.

The genetics of aging is a complex and multifaceted field, and it is likely that we will continue to uncover new and interesting insights into the ways that genes influence longevity.

What happens when DNA becomes damaged?

When DNA becomes damaged, it can lead to a range of consequences, including:

- Mutations: DNA damage can result in mutations, which are changes in the DNA sequence that can affect the function of genes. Mutations can be benign, but they can also lead to changes in the structure and function of proteins, which can have negative consequences for the cell and the organism.

- Cell death: If the damage to DNA is severe, it can lead to cell death. This can occur through a process called apoptosis, which is a programmed form of cell death that occurs naturally in cells when they are damaged beyond repair.

- Senescence: DNA damage can also lead to senescence, which is a state of permanent cell cycle arrest that occurs in response to DNA damage. Senescent cells can no longer divide and reproduce, and they can accumulate with age, contributing to aging and age-related diseases.

- Cancer: DNA damage can increase the risk of cancer by disrupting the normal regulation of cell growth and division. Mutations in genes that regulate cell growth and division can lead to uncontrolled cell growth, which can result in the formation of tumors.

- Aging: DNA damage can also contribute to aging by disrupting the normal functioning of cells and tissues. As cells age, they accumulate damage to their DNA, which can lead to a decline in their function and an increase in their susceptibility to disease.

DNA damage can have significant consequences for cells and organisms, including mutations, cell death, senescence, cancer, and aging. It is important for cells to have mechanisms in place to prevent and repair DNA damage to maintain their health and function over time.

DNA damage is a natural process that occurs in cells over time due to various factors such as environmental stressors, errors during DNA replication, and the natural process of aging. While our cells have powerful mechanisms to repair DNA damage, some damage may remain unrepaired and accumulate over time. This accumulation of DNA damage has been linked to aging and age-related diseases.

There are different types of DNA damage, and their impact on the body varies. For example, small errors in DNA code, called mutations, are generally harmless. However, more severe types of DNA damage, such as a break in a DNA strand, can have more significant consequences.

When a DNA strand breaks, it can be challenging for the body to repair. The repair process is complex, and mistakes can be made, which could shorten lifespan. Therefore, it is essential for cells to have efficient mechanisms to repair DNA damage and maintain genome stability.

Fortunately, our cells have evolved various mechanisms to repair DNA damage. For instance, the body has enzymes that can fix errors in DNA code, and it can also repair breaks in DNA strands. Additionally, cells have a built-in checkpoint system that can halt the cell cycle and prevent damaged cells from dividing until they are repaired.

While DNA damage is a natural part of aging, it is important to note that lifestyle factors such as a healthy diet, regular exercise, and avoiding harmful habits like smoking can help reduce the accumulation of DNA damage and promote health and well-being.

DNA damage is a natural process that occurs in cells over time, and it can accumulate with age. While some damage is harmless, more severe types of damage can have significant consequences. Fortunately, our cells have powerful mechanisms to repair DNA damage, and a healthy lifestyle can help promote genome stability and health.

Telomere shortening is another type of DNA damage that can occur during cell division. Telomeres are the protective caps at the end of chromosomes that prevent the DNA from being damaged or fused with neighboring chromosomes. During cell division, the telomeres shorten, and when they become too short, they can no longer protect the DNA, leaving the cell at risk for serious damage.

In most cells, telomere length cannot be restored, and extreme telomere shortening can trigger an SOS response. The cell will then do one of three things:

  - Stop replicating by turning itself off, becoming senescent.

  - Stop replicating by dying, called apoptosis.

  - Continue to divide, becoming abnormal and potentially dangerous, such as leading to cancer.

The shortening of telomeres is a natural process that occurs with age, and it can contribute to aging and age-related diseases. However, it is also a protective mechanism that prevents damaged cells from dividing and potentially becoming cancerous.

Researchers are exploring ways to lengthen telomeres as a potential anti-aging strategy. One approach is to activate the enzyme telomerase, which can elongate telomeres. However, this approach has its risks, as uncontrolled telomerase activity can lead to cancer.

Therefore, researchers are also exploring ways to modulate telomerase activity to promote healthy aging without increasing the risk of cancer.

Telomere shortening is a natural process that occurs during cell division, and it can contribute to aging and age-related diseases. However, it is also a protective mechanism that prevents damaged cells from dividing and potentially becoming cancerous. Researchers are exploring ways to lengthen telomeres as a potential anti-aging strategy, but it is important to balance this approach with the risk of uncontrolled telomerase activity leading to cancer.

Senescent cells are of great interest to scientists because, despite being in a state of dormancy, they continue to play a significant role in the body's cellular communication and signaling processes. While they are no longer able to divide and proliferate, they remain metabolically active and can interact with other cells through various signaling mechanisms.

One key aspect of senescent cells is their ability to secrete a variety of molecules that can impact the behavior and function of surrounding cells. These molecules, known as senescence-associated secretory phenotype (SASP) factors, can include pro-inflammatory cytokines, chemokines, and growth factors. The release of these molecules can lead to a range of consequences, including the promotion of inflammation, the activation of nearby cells, and the disruption of normal tissue function.

A particular concern is the link between senescent cells and cancer. Senescent cells can contribute to the development and progression of cancer by releasing factors that promote the growth and survival of cancer cells. Additionally, senescent cells can also create a tumor-promoting microenvironment by suppressing the immune response and fostering the growth of cancer stem cells.

Understanding the mechanisms by which senescent cells interact with their surroundings and impact disease processes is a crucial area of research. Scientists are working to develop therapies that target senescent cells and prevent them from releasing harmful molecules, with the goal of reducing the risk of cancer and other age-related diseases.

> **TELOMERE LENGTH: HEALTH SPAN VS. LIFESPAN?**
>
> Aging biologists are investigating whether humans' telomere length is associated with lifespan, health span, or both. In one study of people age 85 years and older, researchers found telomere length was not associated with longevity, at least not in the oldest-old. In another study, researchers analyzing DNA samples from centenarians found that telomeres of healthy centenarians were significantly longer than those of unhealthy centenarians, suggesting that telomere length may be associated with health span.

The relationship between cell senescence, cancer, and aging is complex and multifaceted. Cell senescence, a state in which cells stop dividing and replicating due to damage to their DNA or telomeres, can play a protective role against cancer by preventing severely damaged cells from producing abnormal and potentially cancerous daughter cells. However, as we age, cell senescence can also contribute to the development of cancer by releasing certain molecules that make cells more vulnerable to abnormal function.

Research has shown that senescent cells can secrete a variety of molecules, including pro-inflammatory cytokines, chemokines, and growth factors, that can promote the growth and survival of cancer cells. Additionally, senescent cells can also create a tumor-promoting microenvironment by suppressing the immune response and fostering the growth of cancer stem cells.

Therefore, understanding the mechanisms by which cell senescence contributes to the development of cancer is crucial for the development of effective cancer therapies. Strategies that target senescent cells and prevent them from releasing harmful molecules may help to reduce the risk of cancer and promote healthy aging.

However, it is important to note that cell senescence is not a static process and can be influenced by a variety of factors, including lifestyle choices, environmental exposures, and genetic predisposition. Therefore, a comprehensive approach that takes into account individual differences and factors is necessary for effective cancer prevention and treatment.

Fibroblasts are cells that play a crucial role in maintaining the structure and integrity of tissues in the body, particularly in the skin and other connective tissues. They do this by producing and maintaining the extracellular matrix, a complex network of proteins and other molecules that provides a scaffold for cells to adhere to and communicates with other cells.

Fibroblasts have a limited number of cell divisions, typically around 60, before they enter a state of senescence, which means they stop dividing and proliferating. When fibroblasts turn off, they release molecules that can alter the extracellular matrix and cause inflammation in the surrounding tissue. This can disrupt the normal function of the tissue and contribute to the aging process.

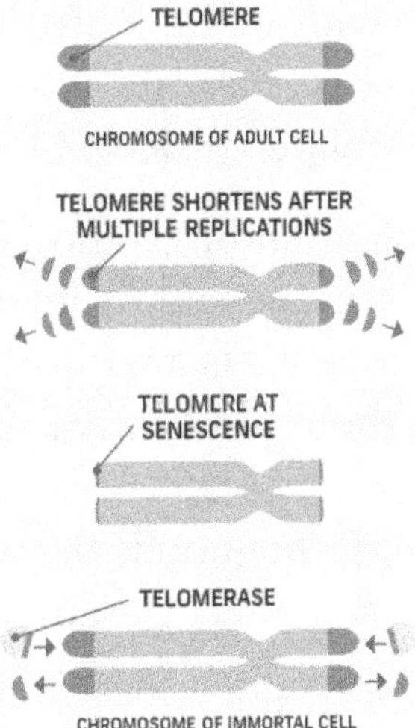

Telomeres shorten each time a cell divides. In most cells, the telomeres eventually reach a critical length when the cells stop proliferating and become senescent.

But, in certain cells, like sperm and egg cells, the enzyme telomerase restores telomeres to the ends of chromosomes. This telomere lengthening insures that the cells can continue to safely divide and multiply. Investigators have shown that telomerase is activated in most immortal cancer cells, since telomeres do not shorten when cancer cells divide.

Additionally, the breakdown of the extracellular matrix can also increase the risk of cancer with age. Cancer cells can use the broken-down extracellular matrix as a scaffold to grow and spread, allowing them to invade and damage surrounding tissues.

YOUNG FIBROBLASTS

OLD FIBROBLASTS, NEARING SENESCENCE

Fibroblasts play a critical role in maintaining tissue structure and integrity, and their senescence can contribute to aging and cancer. Understanding the mechanisms that regulate fibroblast behavior and the breakdown of the extracellular matrix can provide valuable insights into the aging process and the development of age-related diseases.

Understanding the biological mechanisms that govern cell senescence and its impact on aging is a crucial area of research. Cell senescence is a natural process that occurs when cells reach the end of their replicative lifespan or experience stress or damage. Senescent cells can no longer divide and proliferate, but they remain metabolically active and can secrete a variety of molecules that can affect the surrounding cells and tissues.

Early in life, cell senescence plays a beneficial role in preventing cancer and promoting tissue repair. Senescent cells can act as a barrier to prevent damaged or mutated cells from dividing and potentially becoming cancerous. Additionally, senescent cells can secrete factors that promote tissue repair and regeneration.

However, as we age, the accumulation of senescent cells can have detrimental effects. Senescent cells can produce pro-inflammatory cytokines and other molecules that can contribute to chronic inflammation, which is a hallmark of aging. Chronic inflammation can lead to a range of age-related diseases, including cardiovascular disease, arthritis, and neurodegenerative disorders.

Furthermore, senescent cells can also contribute to the breakdown of tissues and organs. As senescent cells accumulate, they can disrupt the normal functioning of tissues and organs, leading to a decline in physical function and an increased risk of age-related diseases.

Researchers are working to understand the underlying mechanisms that govern cell senescence and its impact on aging. By studying the biological pathways that regulate cell senescence, researchers hope to identify potential therapeutic targets that can help to delay or reverse the negative effects of aging.

One promising area of research is the study of senolytic therapies, which are designed to eliminate senescent cells. Senolytic therapies have shown promise in animal models, where they have been shown to improve physical function and reduce the risk of age-related diseases. Researchers are now exploring the use of senolytic therapies in human clinical trials.

Understanding the biological mechanisms that govern cell senescence and its impact on aging is a crucial area of research. By studying the underlying pathways that regulate cell senescence, researchers hope to identify potential therapeutic targets that can help to delay or reverse the negative effects of aging.

METABOLISM

Does stress really shorten your life?

Stress is a natural response of the body to challenging situations, and it is unavoidable in our daily lives. The question remains, however, whether stress truly shortens our lifespan. The answer is not a simple yes or no, as the relationship between stress and aging is complex and multifaceted.

On one hand, chronic stress can have a profound impact on our physical and mental health, increasing the risk of various diseases and accelerating the aging process. When we experience stress, our body's "fight or flight" response is triggered, releasing stress hormones like cortisol and adrenaline into our system. These hormones can lead to inflammation, impaired immune function, and damage to our cardiovascular and metabolic systems.

Chronic stress can also impact our telomeres, which are the protective caps on the ends of our chromosomes. Telomeres naturally shorten as we age, but chronic stress can accelerate this process, leading to premature aging and an increased risk of age-related diseases.

Furthermore, stress can affect our behavior and lifestyle choices, which can also impact our health and aging. For example, people who experience chronic stress may be more likely to engage in unhealthy behaviors like overeating, smoking, or excessive alcohol consumption, all of which can contribute to premature aging.

On the other hand, not all stress is created equal. Some stress, known as "eustress," can actually be beneficial for our health and well-being. Eustress is the positive stress we experience when we are challenged to learn something new, meet new people, or take on new responsibilities. This type of stress can stimulate our brain and body, promoting growth and improvement.

Additionally, some research suggests that the impact of stress on aging may be overstated. A study published in the Journal of Gerontology found that stress levels did not significantly predict mortality or aging-related outcomes in a sample of older adults. Another study published in the journal Psychoneuroendocrinology found that cortisol levels, which are often used as a marker of stress, did not significantly predict aging-related changes in a sample of healthy adults.

It's also important to note that our perception and response to stress can play a significant role in its impact on our health and aging. People who have a positive outlook on life and are able to manage their stress levels through relaxation techniques, social support, and healthy lifestyle choices may be less likely to experience the negative effects of stress on their health and aging.

While chronic stress can have a negative impact on our health and aging, not all stress is created equal. Eustress, or positive stress, can actually be beneficial for our growth and well-being. Additionally, our perception and response to stress can play a significant role in its impact on our health and aging. It's important to manage stress levels through healthy lifestyle choices, relaxation techniques, and social support to minimize its negative effects and promote healthy aging.

Metabolism is a vital process that occurs in the body and is essential for life. It is the process by which the body converts food and oxygen into energy, which is then used to power all of the body's various functions. These functions include breathing, circulating blood, eliminating waste, controlling body temperature, contracting muscles, operating the brain and nerves, and just about every other activity associated with living.

When we eat food, our body breaks it down into glucose, which is then used by our cells to produce energy. This process is called cellular respiration. The energy produced from cellular respiration is then used to power all of the body's functions.

In addition to producing energy, metabolism also plays a role in maintaining homeostasis in the body. Homeostasis is the ability of the body to maintain a stable internal environment despite changes in external conditions. This is achieved through a complex network of hormones and other chemical messengers that work together to regulate various bodily functions.

Metabolism can be influenced by a number of factors, including genetics, lifestyle, and diet. For example, some people may have a faster metabolism than others due to their genetic makeup, which can affect their ability to lose weight or maintain weight loss. Lifestyle factors such as physical activity level and diet can also impact metabolism. A healthy diet that includes plenty of fruits, vegetables, whole grains, and lean protein sources can help support a healthy metabolism, while a sedentary lifestyle can slow down metabolism and contribute to weight gain.

Metabolism is a vital process that occurs in the body and is essential for life. It is the process by which the body converts food and oxygen into energy, which is then used to power all of the body's various functions. Factors such as genetics, lifestyle, and diet can influence metabolism, and maintaining a healthy metabolism is important for health and well-being.

The statement "These everyday metabolic activities that sustain life also create 'metabolic stress,' which, over time, results in damage to our bodies" is a profound observation that highlights the double-edged nature of metabolism. On the one hand, metabolism is essential for sustaining life by providing the energy and molecules needed for various physiological processes. On the other hand, the very same metabolic processes that sustain life also create stress on the body, which can lead to damage and dysfunction over time.

The example of breathing and the role of oxygen in creating metabolic stress is particularly insightful. Oxygen is essential for the production of energy in cells, but it also has a dark side. The metabolism of oxygen produces potentially harmful byproducts called oxygen free radicals, which can react with and damage surrounding molecules, leading to a chain reaction of oxidation. This process can ultimately result in the breakdown or rearrangement of molecules, leading to cellular damage and dysfunction.

The production of oxygen free radicals is not limited to breathing; it can also be triggered by various environmental factors, such as exposure to tobacco smoke and sunlight. This means that even seemingly innocuous activities like smoking or spending time outdoors can contribute to metabolic stress and damage to the body.

The concept of metabolic stress highlights the importance of balance in metabolic processes. While metabolism is essential for sustaining life, an imbalance in metabolic processes can lead to damage and dysfunction. This is why it is important to maintain a healthy lifestyle, including a balanced diet, regular exercise, and adequate sleep, to minimize the negative effects of metabolic stress.

Furthermore, the discussion of metabolic stress also underscores the importance of antioxidants in maintaining cellular health. Antioxidants are molecules that can neutralize oxygen free radicals, preventing them from damaging surrounding molecules and triggering a chain reaction of oxidation. Consuming a diet rich in antioxidants, such as fruits, vegetables, and nuts, can help mitigate the negative effects of metabolic stress and promote cellular health.

The statement "These everyday metabolic activities that sustain life also create 'metabolic stress,' which, over time, results in damage to our bodies" is a powerful reminder of the complex interplay between metabolism and cellular health. While metabolism is essential for sustaining life, it is important to be mindful of the potential negative effects of metabolic stress and take steps to minimize them through a healthy lifestyle and a balanced diet rich in antioxidants.

Free Radicals: Both Friend and Foe

Free radicals, unstable molecules with unpaired electrons, have a dual nature. While some free radicals are beneficial, such as those used by the immune system to destroy harmful organisms, others can cause damage to molecules, including proteins and DNA. This damage can lead to a decline in mitochondrial function, resulting in less energy production and an increase in free radicals. Understanding the role of free radicals in the body is crucial for maintaining cellular health and preventing disease.

Beneficial Free Radicals

Oxygen free radicals, also known as reactive oxygen species (ROS), play a vital role in the immune system. White blood cells, such as neutrophils and macrophages, use ROS to destroy bacteria and other harmful organisms. ROS are also involved in the production of cytokines, signaling molecules that help coordinate the immune response.

In addition, oxidation and its by-products are essential for nerve cell communication in the brain. Nerve cells use ROS to activate signaling pathways, which allows them to communicate with each other.

Harmful Free Radicals

However, not all free radicals are beneficial. In fact, most free radicals are harmful and can cause damage to molecules, including proteins and DNA. This damage can accumulate over time, leading to cellular aging and dysfunction.

Mitochondria, the powerhouses of the cell, are particularly susceptible to free radical damage. Mitochondria metabolize oxygen to produce energy, which generates ROS as a byproduct. As damage mounts, mitochondria may become less efficient, producing less energy and more free radicals. This cycle can lead to a decline in cellular function, contributing to aging and age-related diseases.

HEAT SHOCK PROTEINS

In the early 1960s, scientists discovered that fruit flies exposed to a burst of heat produced proteins that helped their cells survive the temperature change. Over the years, scientists have found these "heat shock proteins" in virtually every living organism, including plants, bacteria, worms, mice, and even humans. Scientists have learned that, despite their name, heat shock proteins are produced when cells are exposed to a variety of stresses, not just heat. The proteins can be triggered by oxidative stress and by exposure to toxic substances (for example, some chemicals). When heat shock proteins are produced, they help cells dismantle and dispose of damaged proteins and help other proteins keep their structure and not become unraveled by stress. They also facilitate making and transporting new proteins in the body.

Heat shock response to stress changes with age. Older animals have a higher everyday level of heat shock proteins, indicating that their bodies are under more biological stress than younger animals. On the other hand, older animals are unable to produce an adequate amount of heat shock proteins to cope with fleeting bouts of stress from the environment.

Heat shock proteins are being considered as a possible aging biomarker—something that could predict lifespan or development of age-related problems—in animal models like worms and fruit flies. However, the exact role heat shock proteins play in the human aging process is not yet clear.

Antioxidants to the Rescue

Fortunately, there are antioxidants, molecules that can neutralize free radicals and prevent them from causing damage. Antioxidants, such as vitamins C and E, can donate electrons to free radicals, stabilizing them and preventing them from reacting with other molecules.

Antioxidants can be found in a variety of foods, including fruits, vegetables, nuts, and grains. Consuming a diet rich in antioxidants can help protect cells from free radical damage and promote cellular health.

Free radicals are both beneficial and harmful, depending on their context. While some free radicals play a vital role in the immune system and nerve cell communication, others can cause damage to molecules and contribute to cellular aging. Understanding the role of free radicals in the body is essential for maintaining cellular health and preventing disease. By consuming a diet rich in antioxidants, we can help protect cells from free radical damage and promote cellular health.

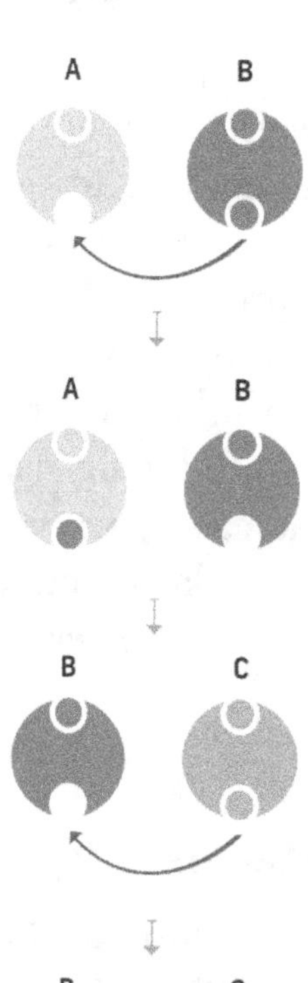

Oxidative stress, which occurs when there is an imbalance between the production of free radicals and the body's ability to neutralize them, has been implicated in the aging process and a variety of age-related diseases. Free radicals, also known as reactive oxygen species (ROS), are unstable molecules that can damage cellular components, including DNA, proteins, and lipids.

The accumulation of oxidative damage over time has been linked to many age-related changes, such as cellular senescence, inflammation, and the degradation of cellular components. For example, telomeres, the protective caps at the end of chromosomes, can become shortened and damaged due to oxidative stress, leading to cellular aging.

Free radicals have also been implicated in various age-related diseases, including:

- Cancer: Oxidative stress can lead to DNA damage, which can increase the risk of cancer.

- Atherosclerosis: Oxidative stress can damage blood vessels, leading to the formation of plaques and increasing the risk of heart disease.

- Cataracts: Oxidative stress can damage the lens of the eye, leading to clouding and impaired vision.

- Neurodegeneration: Oxidative stress can damage brain cells, leading to neurodegenerative diseases such as Alzheimer's and Parkinson's.

While the exact mechanisms by which oxidative stress contributes to aging and age-related diseases are not yet fully understood, there is evidence to suggest that reducing oxidative stress may help to promote healthy aging and reduce the risk of age-related diseases. This can be achieved through a combination of dietary and lifestyle modifications, such as consuming antioxidant-rich foods, exercising regularly, and avoiding smoking and excessive alcohol consumption.

Antioxidants are substances that help protect cells from damage caused by free radicals. They can neutralize free radicals by donating an electron to them, which stabilizes the molecule and prevents it from reacting with other cellular components.

Antioxidants can be found in many foods, such as fruits, vegetables, nuts, and grains. They are also produced naturally in the body through various enzymatic processes. Some examples of antioxidants include:

- Vitamins C and E: These are two of the most well-known antioxidants. Vitamin C is a powerful antioxidant that can neutralize a wide range of free radicals, while vitamin E is particularly effective against free radicals that are formed during the metabolism of fat.

- Superoxide dismutase (SOD): This enzyme is produced naturally in the body and is responsible for neutralizing superoxide radicals, which are formed when cells use oxygen to produce energy.

- Catalase: This enzyme is also produced naturally in the body and is responsible for neutralizing hydrogen peroxide, which is a byproduct of oxidative metabolism.

- Glutathione peroxidase: This enzyme is produced naturally in the body and is responsible for neutralizing hydrogen peroxide and other peroxides.

Antioxidants play a crucial role in protecting cells from damage caused by free radicals. By neutralizing these harmful molecules, antioxidants help to maintain cellular health and prevent damage that can lead to chronic diseases such as cancer, heart disease, and neurodegenerative disorders.

The concept of antioxidants and their role in countering the negative effects of oxygen free radicals has been extensively studied in various animal models. While some experiments have shown promising results, others have yielded conflicting outcomes.

One of the most well-known studies was conducted by NIA-supported researchers, who inserted extra copies of the SOD gene into fruit flies, resulting in a 30% increase in their average lifespan. Similarly, another study found that immersing roundworms in a synthetic form of SOD and catalase extended their lifespan by 44%.

However, these results were not consistently replicated in other studies. For example, a comprehensive set of experiments involving laboratory mice found that increasing or decreasing antioxidant enzymes had no significant effect on lifespan.

Human clinical trials involving antioxidants have also yielded mixed results. While some studies have suggested that antioxidant supplementation may have health benefits, such as reducing the risk of certain diseases, others have found no significant effect on lifespan.

The relationship between antioxidants and lifespan remains a topic of ongoing investigation and debate. While some studies suggest that antioxidants may have a positive impact on lifespan, others have found no conclusive evidence to support this theory. Further research is needed to fully understand the role of antioxidants in aging and longevity.

Does how much you eat affect how long you live?

Yes, how much you eat can affect how long you live. Overeating can lead to a shorter lifespan and serious health problems, such as cardiovascular disease and type 2 diabetes, due to the added stress placed on the body. The body needs food to survive, but the process of metabolizing food can create stress on the body. Overeating can lead to even more stress, which can negatively impact health and lifespan.

When we eat, our body uses the energy from the food to fuel its various functions, such as breathing, digestion, and physical activity. However, the process of metabolizing food also creates stress on the body. This stress can come from a variety of sources, including:

- Oxidative stress: When we eat, our body produces free radicals, which are unstable molecules that can damage our cells and contribute to aging and disease.

- Inflammation: Consuming high amounts of sugar, refined carbohydrates, and unhealthy fats can lead to chronic inflammation, which can increase the risk of diseases such as heart disease, cancer, and Alzheimer's.

- Insulin resistance: Overeating can lead to insulin resistance, a condition in which the body becomes less responsive to insulin, leading to high blood sugar levels and an increased risk of developing type 2 diabetes.

- Metabolic syndrome: Overeating can also contribute to metabolic syndrome, a cluster of conditions that increase the risk of developing heart disease, stroke, and diabetes.

When we overeat, we are placing additional stress on our body, which can lead to a range of negative health effects. These effects can include:

- Weight gain: Overeating can lead to weight gain, which can increase the risk of developing a range of health problems, including heart disease, stroke, and type 2 diabetes.

- Increased risk of chronic diseases: Overeating can increase the risk of developing chronic diseases such as heart disease, stroke, and type 2 diabetes.

- Reduced lifespan: Consuming high amounts of food can lead to a shorter lifespan, as the added stress on the body can increase the risk of developing age-related diseases.

It's important to note that not all foods are created equal, and some foods are better for our health than others. Consuming a diet rich in whole, unprocessed foods, such as fruits, vegetables, whole grains, and lean proteins, can help to minimize the negative effects of overeating and promote health and well-being.

Overeating can have negative effects on our health and lifespan, and it's important to be mindful of our food choices and consume a balanced diet that supports our health and well-being.

Calorie restriction is a dietary approach that has been shown to increase lifespan and delay the onset of age-related diseases in various animal models. It involves reducing calorie intake by at least 30% compared to a normal diet, while still maintaining a balanced intake of protein, fat, vitamins, and minerals.

The concept of calorie restriction was first introduced in the 1930s, when researchers found that laboratory rats and mice lived up to 40% longer when fed a calorie-restricted diet. Since then, numerous studies have been conducted on various animal models, including yeast, worms, flies, some strains of mice, and nonhuman primates.

The results of these studies have consistently shown that calorie restriction increases lifespan and delays the onset of age-related diseases, such as cancer, cardiovascular disease, and neurodegenerative disorders. Additionally, calorie restriction has been found to delay normal age-related decline in physical function and cognitive function.

It is important to note that calorie restriction is not exclusive to a research setting and can be applied in a healthy eating or dieting context. However, it is a more tightly controlled approach than traditional healthy eating or dieting, as it involves a significant reduction in calorie intake and a focus on maintaining a balanced nutrient intake.

While the exact mechanisms by which calorie restriction works are not fully understood, it is believed that it activates cellular pathways that promote longevity and reduce oxidative stress, inflammation, and other age-related processes.

Calorie restriction is a promising approach for increasing lifespan and health span, and its effects have been consistently demonstrated in a variety of animal models. However, it is important to note that more research is needed to fully understand its effects in humans and to determine the optimal way to implement calorie restriction in a healthy eating or dieting context.

Calorie restriction has been shown to have positive effects on age-related diseases and mortality in nonhuman primates. Two studies, one conducted at the National Institute on Aging (NIA) and the other supported by NIA at the University of Wisconsin, have demonstrated that calorie-restricted diets in monkeys can lead to a decrease and/or delay in the onset of age-related diseases, as well as a reduction in age-related deaths.

In the NIA study, monkeys fed a calorie-restricted diet had a notably decreased and/or delayed onset of age-related diseases compared to the control group of "normal" eaters. The study found that the calorie-restricted monkeys had fewer age-related diseases, such as cancer, cardiovascular disease, and kidney disease, compared to the control group.

The University of Wisconsin study also found that calorie-restricted rhesus monkeys had three times fewer age-related diseases compared to the control group. In addition, the study found that the calorie-restricted monkeys had fewer age-related deaths compared to their normal-fed controls.

While the findings from the NIA study were published in 2007, it was too early to determine whether calorie restriction had any effects on lifespan. However, research in primates continues, and further studies are needed to fully understand the effects of calorie restriction on aging and lifespan in nonhuman primates.

The results of these studies suggest that calorie restriction may have a positive impact on age-related diseases and mortality in nonhuman primates, and provide support for further research into the potential benefits of calorie restriction in humans.

Calorie restriction, while shown to increase lifespan in some animal models, does not have a uniform effect across all species. Studies on wild mice have revealed that calorie restriction does not necessarily lead to increased lifespan, and in some cases, may even shorten lifespan. This variability may be attributed to genetic differences between species or strains.

A 2010 study funded by the National Institute on Aging (NIA) further supports the role of genetics in determining the effectiveness of calorie restriction on longevity. The study examined 42 closely related strains of laboratory mice and found that only about a third of the strains on a calorie-restricted diet had an increase in longevity. One-third of the strains had a shortened lifespan, while the remaining third showed no significant difference in lifespan compared to mice on a normal diet.

These findings suggest that genetics play a significant role in determining whether calorie restriction will have a positive effect on longevity. While calorie restriction may be beneficial for some individuals, it is not a universal solution for increasing lifespan and may have varying effects depending on the species or strain.

Calorie restriction has been shown to have positive effects on various health markers in animal studies. To determine if these effects can be replicated in humans, researchers have conducted pilot studies and are currently conducting a longer-term trial.

The Comprehensive Assessment of Long-term Effects of Reducing Intake of Energy (CALERIE) study, conducted in 2002, was a pilot study that aimed to assess the safety and practicality of calorie restriction in humans. The study involved reducing daily caloric intake by 25% for a period of one year.

The results of the study showed that participants had lowered their fasting glucose, total cholesterol, core body temperature, body weight, and fat after one year. Additionally, at the cellular level, participants had better functioning mitochondria and reduced DNA damage.

However, the study also revealed that adapting and adhering to the calorie restriction regimen could be difficult. Participants reported feeling hungry and experiencing a decrease in their quality of life.

Despite these challenges, a longer-term trial is currently underway to further investigate the effects of calorie restriction on human health. The trial, known as the CALERIE-2 study, aims to assess the safety and efficacy of calorie restriction over a period of two years.

While the results of the CALERIE-2 study are still pending, the findings from the pilot study suggest that calorie restriction may have positive effects on various health markers in humans. However, more research is needed to determine the long-term safety and practicality of this approach.

Potential research questions related to the mechanisms and pathways of calorie restriction:

- What are the key cellular and molecular mechanisms involved in the anti-aging effects of calorie restriction?

- How does calorie restriction affect the regulation of cellular metabolism, energy production, and oxidative stress?

- What are the roles of various hormesis pathways, such as the AMP-activated protein kinase (AMPK) and sirtuin 1 (SIRT1) pathways, in mediating the beneficial effects of calorie restriction?

- How does calorie restriction impact the activity of various cellular compartments, such as mitochondria and lysosomes, and what are the implications for cellular health and function?

- Can calorie restriction induce a state of "cellular longevity" or "cellular rejuvenation," and if so, what are the underlying mechanisms?

- How do the benefits of calorie restriction compare to those of other interventions, such as exercise and metformin, and what are the potential synergies or conflicts between these approaches?

- Can calorie restriction have a positive impact on various age-related diseases, such as cancer, cardiovascular disease, and neurodegenerative disorders, and if so, what are the underlying mechanisms?

- What are the potential side effects and risks of calorie restriction, and how can they be minimized or mitigated?

- Can calorie restriction be combined with other interventions, such as drug therapies or supplements, to enhance its anti-aging effects, and if so, what are the potential benefits and risks of these combinations?

- How can we translate the findings from animal studies of calorie restriction to human populations, and what are the potential challenges and limitations of this process?

Regarding compounds that might mimic the benefits of calorie restriction, some potential research questions could include:

- Can we identify small molecules or nutrients that activate the same cellular pathways as calorie restriction, such as AMPK or SIRT1, and if so, what are their potential benefits and risks?

- Can we develop drugs or supplements that target specific cellular mechanisms involved in the anti-aging effects of calorie restriction, such as mitochondrial function or oxidative stress?

- Can we create a "calorie restriction mimetic" that recapitulates the beneficial effects of calorie restriction without the need for actual calorie restriction?

- How do the benefits and risks of calorie restriction mimetics compare to those of other anti-aging interventions, such as metformin or rapamycin?

- Can we use machine learning or artificial intelligence algorithms to identify potential calorie restriction mimetics and optimize their design and development?

These are just a few examples of potential research questions related to calorie restriction and its mechanisms, and there are likely many other interesting and important questions to explore in this area.

The mechanisms by which calorie restriction (CR) exerts its anti-aging effects are not yet fully understood, but a wide range of possibilities are being investigated. One theory is that reducing caloric intake leads to a decrease in oxidative damage to cells, which can contribute to aging. Another hypothesis suggests that the relative scarcity of nutrients caused by CR triggers the body's defense mechanisms, including heat shock proteins, which help the body to better withstand other stresses and health problems.

Some researchers believe that the effects of CR are controlled by the brain and nervous system. Studies have shown that CR can increase the production of brain-derived neurotrophic factor (BDNF), a protein that protects the brain from dysfunction and degeneration, and supports increased regulation of blood sugar and heart function in animal models.

Other potential mechanisms by which CR may work include its influence on hormonal balance, cell senescence, or gene expression. It is likely that CR acts through a combination of these mechanisms, and others yet to be identified.

The study by the National Institute on Aging (NIA) mentioned in the passage supports the idea that CR can increase BDNF production, which in turn can protect the brain from dysfunction and degeneration. This study provides evidence that CR can have anti-aging effects on the brain and nervous system, and suggests that BDNF may play a key role in these effects.

The mechanisms by which CR works are complex and multifaceted, and likely involve a combination of cellular, hormonal, and neural processes. Further research is needed to fully understand how CR exerts its anti-aging effects, and to identify potential therapeutic targets for age-related diseases.

The sirtuins and mTOR pathways play a crucial role in regulating cellular metabolism, energy homeostasis, and cell survival. Both pathways have been implicated in the aging process, and their dysregulation has been linked to various age-related diseases. Calorie restriction, a dietary intervention that involves reducing daily caloric intake, has been shown to activate these pathways and promote longevity in various species.

Sirtuins are a family of NAD+-dependent deacetylases that play a key role in regulating cellular metabolism, energy homeostasis, and cell survival. There are seven known mammalian sirtuins (SIRT1-7), each with distinct functions and tissue-specific expressions. Sirtuins regulate various cellular processes, including DNA repair, cell cycle progression, metabolism, and stress resistance.

The mTOR pathway is a key regulator of cellular metabolism and growth. It integrates signals from nutrient availability, growth factors, and energy status to regulate cell growth, proliferation, and autophagy. mTOR is activated in response to nutrient availability and growth factors, and it regulates cell growth by phosphorylating and activating key downstream effectors, including the translation initiation factor 4E-BP1 and the cell cycle regulator S6K -

Calorie restriction has been shown to activate the sirtuins and mTOR pathways in various species. SIRT1 has been shown to be activated in response to calorie restriction in various tissues, including the brain, muscle, and liver. Activation of SIRT1 has been linked to improved insulin sensitivity, reduced inflammation, and enhanced stress resistance.

Rapamycin, an mTOR inhibitor, has also been shown to mimic the effects of calorie restriction in various species. Rapamycin activates autophagy, a cellular recycling process that helps maintain cellular homeostasis and promote cell survival. Rapamycin has been shown to promote longevity in mice, and it is being explored as a potential therapeutic agent for age-related diseases.

Resveratrol, a polyphenol found in grapes and other plant-based foods, has also been shown to activate the sirtuins and mTOR pathways. Resveratrol has been shown to promote longevity in various species, and it is being explored as a potential therapeutic agent for age-related diseases.

The sirtuins and mTOR pathways play a crucial role in regulating cellular metabolism, energy homeostasis, and cell survival. Calorie restriction, resveratrol, and rapamycin have been shown to activate these pathways and promote longevity in various species. Understanding the molecular mechanisms underlying these pathways may provide valuable insights into the aging process and the development of therapeutic interventions for age-related diseases.

Resveratrol, a polyphenol found in grapes, wine, and nuts, has been shown to activate the sirtuin pathway and have anti-aging effects in various species. Studies have demonstrated that resveratrol can increase the lifespan of yeast, flies, worms, and fish. In mice, resveratrol has been shown to lessen the negative effects of a high-fat and high-calorie diet, improve health, and prevent age- and obesity-related decline in heart function. It has also been found to partially mimic the effects of calorie restriction on gene expression profiles in liver, skeletal muscle, and adipose tissue. However, resveratrol did not have an impact on the mice's survival or maximum lifespan. These findings suggest that resveratrol may not affect all aspects of the basic aging process and that there may be different mechanisms for health versus lifespan.

Research on resveratrol continues in mice, as well as in nonhuman primates and people. Scientists are interested in exploring the potential anti-aging effects of resveratrol in humans, as well as its potential to prevent or treat age-related diseases. However, it is important to note that more research is needed to fully understand the effects of resveratrol on aging and health in humans.

One question that remains to be answered is why resveratrol did not have an impact on the mice's survival or maximum lifespan. It is possible that resveratrol may have different effects on different species or that the dosage used in the study was not sufficient to have an impact on lifespan. Additionally, it is possible that resveratrol may have different effects on health and lifespan in humans.

Further research is needed to fully understand the effects of resveratrol on aging and health in humans. This could include studies that examine the impact of resveratrol on various age-related diseases, such as heart disease, Alzheimer's disease, and cancer. Additionally, researchers could explore the optimal dosage and duration of resveratrol supplementation for anti-aging effects.

Resveratrol has shown promise as an anti-aging compound in animal studies, but more research is needed to fully understand its effects in humans. While it may not have an impact on survival or maximum lifespan, it may still have anti-aging effects on health and well-being.

Rapamycin is a compound that has gained attention for its potential to mimic the effects of calorie restriction and extend lifespan. It works by inhibiting the mTOR pathway, which is a key regulator of cell growth and metabolism. The mTOR pathway is activated in response to nutrient availability and growth factors, and it plays a critical role in regulating cell growth, proliferation, and autophagy.

Rapamycin has been shown to extend the median and maximum lifespan of mice in several studies. For example, a study published in 2009 by the NIA's Interventions Testing Program found that rapamycin extended the median and maximum lifespan of mice by 13% and 17%, respectively. This study also found that rapamycin had a positive effect on the mice's healthspan, as measured by their ability to perform physical tasks and their cognitive function.

One of the key advantages of rapamycin is that it can be administered later in life and still have a positive effect on lifespan. In the study mentioned above, rapamycin was fed to the mice beginning at early-old age (20 months), and it still had a significant impact on their lifespan. This suggests that an intervention started later in life may still be able to increase longevity, which is an important consideration for people who may not have started their anti-aging regimen until later in life.

Researchers are now exploring the effects of rapamycin on health span and looking for other compounds that may have similar effects on the mTOR pathway. For example, a study published in 2017 found that a compound called AZD-8055, which inhibits the mTOR pathway, extended the lifespan of mice by 14%. Another study published in 2020 found that a combination of rapamycin and a compound called NAD+, which activates the mTOR pathway, extended the lifespan of mice by 22%.

While these studies are promising, it's important to note that more research is needed to fully understand the effects of rapamycin and other mTOR inhibitors on human lifespan and healthspan. Additionally, it's important to consult with a healthcare professional before starting any new supplement or medication, especially if you have any pre-existing health conditions or are taking medications that may interact with rapamycin.

Rapamycin is a compound that has shown promise in extending lifespan and improving healthspan by inhibiting the mTOR pathway. While more research is needed, it's an exciting area of study that may ultimately lead to new anti-aging therapies for humans.

It is fascinating to consider the potential of calorie restriction mimetics, such as resveratrol and rapamycin, to influence human aging. While the effects of these compounds on human aging are not yet fully understood, research in this area may uncover valuable insights into the aging process and potentially lead to the development of new anti-aging therapies.

One key area of investigation is the examination of the mechanisms and pathways underlying calorie restriction. By understanding how calorie restriction affects the body's metabolism, scientists may be able to identify specific targets for intervention and develop therapies that can mimic the beneficial effects of calorie restriction without requiring significant dietary changes.

Resveratrol, for example, has been shown to activate SIRT1, a protein that plays a role in maintaining cellular health and preventing age-related diseases. Rapamycin, on the other hand, inhibits the mTOR pathway, which is involved in cell growth and metabolism. Understanding how these compounds interact with these pathways and how they affect the aging process may lead to the development of new therapies that can promote healthy aging.

Another area of investigation is the study of the effects of calorie restriction mimetics on various age-related diseases. For example, research has shown that resveratrol may have a protective effect against neurodegenerative diseases such as Alzheimer's and Parkinson's, while rapamycin has been shown to have anti-cancer properties. By exploring the effects of these compounds on different age-related diseases, scientists may be able to identify new ways to prevent or treat these conditions.

While the potential of calorie restriction mimetics is promising, it is important to note that much work remains to be done. More research is needed to fully understand the effects of these compounds on human aging and to identify safe and effective dosages for use in humans. Additionally, it is likely that different individuals may respond differently to these compounds, so personalized treatment plans may be necessary.

Learning more about calorie restriction mimetics and the mechanisms and pathways underlying calorie restriction may point the way to future healthy aging therapies. While there is still much to be discovered, the potential of these compounds to promote healthy aging is an exciting area of research that may lead to new breakthroughs in the years to come.

IMMUNE SYSTEM

Can your immune system still defend you as you age?

Yes, the immune system can still defend the body against infections and diseases as people age, but it may not be as effective as it was in younger years. This is because the immune system tends to decline with age, making older adults more susceptible to infections and less able to fight them off quickly.

The immune system is complex and has many different components that work together to protect the body. One of the key components is the T-cell, which is responsible for recognizing and attacking infected cells. As people age, the number and function of T-cells can decrease, making it harder for the immune system to fight off infections.

Another important component of the immune system is the antibody, which is a protein that recognizes and attacks specific pathogens. Antibodies are produced by the immune system in response to exposure to a specific pathogen, and they can provide long-term protection against future infections. However, the production of antibodies can decrease with age, making it harder for the immune system to mount a strong response to new infections.

There are several factors that can contribute to the decline of the immune system with age. One factor is the natural aging process, which can lead to a decrease in the number and function of immune cells. Another factor is lifestyle, as people who lead a sedentary lifestyle, have a poor diet, or smoke may have a weaker immune system. Chronic diseases such as diabetes, heart disease, and arthritis can also weaken the immune system.

Despite the decline of the immune system with age, there are steps that older adults can take to keep their immune system strong. These include:

* Getting enough sleep

* Exercising regularly

* Eating a healthy diet rich in fruits, vegetables, and whole grains

* Staying up to date on vaccines

* Practicing good hygiene, such as washing hands frequently and avoiding close contact with people who are sick

* Avoiding smoking and limiting alcohol consumption

By taking these steps, older adults can help keep their immune system strong and reduce their risk of infections and diseases.

The immune system is indeed a complex network of cells, tissues, and organs that work together to protect the body against infection and disease. It is composed of two major parts: the innate immune system and the adaptive immune system.

The innate immune system is the first line of defense against pathogens and is activated immediately upon infection. It includes physical barriers such as skin and mucous membranes, as well as cells such as neutrophils and macrophages that can recognize and attack invading pathogens. The innate immune system is also responsible for activating the adaptive immune system.

The adaptive immune system, on the other hand, is a specific response to a pathogen that takes several days to develop. It involves the activation of T-cells and B-cells, which recognize and attack specific pathogens. The adaptive immune system also has a memory component, which allows it to remember specific pathogens and mount a stronger response upon future exposure.

As people age, both the innate and adaptive immune systems undergo changes that can affect their ability to fight off infection and disease. For example, the innate immune system may become less effective at recognizing and attacking pathogens, while the adaptive immune system may have a harder time mounting a strong response.

Studies have shown that certain age-related changes in the immune system can increase the risk of infection and disease. For example, older adults may have a harder time fighting off viral infections, such as the flu, and may be more susceptible to certain types of cancer.

Researchers are working to better understand the changes that occur in the immune system with age and to identify ways to support the aging immune system. Some studies have shown that certain lifestyle factors, such as exercise and a healthy diet, can help to boost the immune system. Other studies are exploring the use of immunomodulatory drugs and vaccines to support the immune system.

It is important to understand the changes that occur in the immune system with age and to take steps to support immune function. By doing so, we can help to protect ourselves against infection and disease and maintain optimal health as we age.

Innate immunity is indeed our first line of defense against harmful pathogens, and it consists of various physical barriers and specialized cells that work together to prevent the entry of pathogens into the body. The skin, mucous membranes, cough reflex, and stomach acid are all important components of innate immunity.

As you mentioned, research has shown that with age, innate immune cells can lose some of their ability to communicate with each other effectively. This can make it more difficult for the immune system to respond adequately to pathogens, including viruses and bacteria. This decline in innate immune function is a natural part of the aging process, but it can be exacerbated by various factors such as lifestyle choices, environmental exposures, and genetic predisposition.

Fortunately, there are ways to support innate immunity and reduce the risk of infection and disease. For example, maintaining good hygiene practices, such as washing hands regularly and avoiding close contact with people who are sick, can help prevent the spread of pathogens. Additionally, a healthy diet rich in fruits, vegetables, and whole grains can provide essential nutrients and antioxidants that support immune function.

Exercise, stress management, and adequate sleep are also important for maintaining a healthy immune system. Vaccines and immunomodulatory drugs can also be used to support innate immunity and help protect against infection and disease.

It's important to take a comprehensive approach to supporting immune function, including both innate and adaptive immunity, to maintain optimal health throughout the aging process.

Inflammation is a natural response of the immune system to injury or infection, and it plays a crucial role in the body's defense against pathogens. In young individuals, acute inflammation is a vital part of the immune system's defense against disease, and it helps to initiate the healing process. However, as people age, they tend to experience mild, chronic inflammation, which is associated with an increased risk of various diseases, such as heart disease, arthritis, frailty, type 2 diabetes, physical disability, and dementia.

Researchers are still investigating the relationship between inflammation and disease in older adults. It is not clear whether inflammation leads to disease, disease leads to inflammation, or both scenarios are true. However, studies have shown that centenarians and other older individuals who have maintained good health tend to have less inflammation and a more efficient recovery from infection and inflammation compared to those who are unhealthy or have average health.

The underlying causes of chronic inflammation in older individuals are complex and multifaceted. Some contributing factors include:

- Aging-related changes in the immune system: As people age, their immune system undergoes changes that can lead to chronic inflammation. For example, the thymus gland, which produces T-cells, shrinks with age, leading to a decline in the number and function of T-cells. This can make it harder for the body to fight off infections and can lead to chronic inflammation.

- Oxidative stress: Oxidative stress occurs when the body produces more free radicals than it can neutralize. Free radicals are unstable molecules that can damage cells and tissues, leading to inflammation. As people age, their bodies produce fewer antioxidants, which can lead to oxidative stress and chronic inflammation.

- Lifestyle factors: Lifestyle factors such as smoking, lack of exercise, and poor diet can contribute to chronic inflammation.

Smoking, for example, can damage the lungs and lead to chronic inflammation, while a lack of exercise and poor diet can lead to obesity, which is a major risk factor for chronic inflammation.

- Chronic diseases: Chronic diseases such as arthritis, diabetes, and heart disease can all contribute to chronic inflammation. For example, in arthritis, the joints become inflamed, leading to pain and stiffness. In diabetes, high blood sugar levels can lead to chronic inflammation, which can damage blood vessels and nerves.

Understanding the underlying causes of chronic inflammation in older individuals is important because it can help gerontologists find ways to temper its associated diseases. For example, lifestyle changes such as quitting smoking, exercising regularly, and eating a healthy diet can help to reduce chronic inflammation. In addition, medications such as anti-inflammatory drugs and corticosteroids can be used to reduce inflammation in some cases.

Chronic inflammation is a major concern for older adults, as it is associated with an increased risk of various diseases. Understanding the underlying causes of chronic inflammation and how it relates to disease is crucial for developing effective strategies to promote healthy aging and reduce the risk of age-related diseases. By studying centenarians and other older individuals who have maintained good health, researchers may be able to identify factors that can help to temper chronic inflammation and promote healthy aging.

The adaptive immune system is indeed more complex than the innate immune system and consists of various organs and tissues that work together to provide a specific response to pathogens. The thymus, spleen, tonsils, bone marrow, circulatory system, and lymphatic system are all involved in the adaptive immune response.

T cells, also known as lymphocytes, are a type of white blood cell that plays a central role in the adaptive immune response. They are responsible for recognizing and eliminating infected cells, viruses, and other foreign substances from the body. T cells are produced in the bone marrow and mature in the thymus, where they become functional and can recognize and respond to specific antigens.

There are several different types of T cells, each with its own unique function. CD4+ T cells, also known as helper T cells, play a critical role in activating other immune cells, such as B cells and macrophages, to fight infection. CD8+ T cells, also known as cytotoxic T cells, directly kill infected cells or tumor cells.

Gerontologists are interested in T cells because their function declines with age, which can leave older adults more susceptible to infections and less able to fight off diseases. Studies have shown that the thymus, which is responsible for producing and maturing T cells, begins to shrink and become less functional with age. This can lead to a decrease in the number and diversity of T cells, making it more difficult for the immune system to respond effectively to new infections.

In addition, older adults may have a harder time producing new T cells in response to infection, which can further compromise their immune function. This is why vaccines, which rely on the immune system to produce antibodies to protect against infection, may be less effective in older adults.

Understanding how the adaptive immune system, including T cells, functions and changes with age is crucial for developing effective strategies to promote healthy aging and prevent age-related diseases.

T cells, also known as T lymphocytes, play a crucial role in cell-mediated immunity, which is the immune system's defense against infected or damaged cells. They can directly attack infected cells or produce chemical signals that activate other immune cells and substances.

Before a T cell can recognize and fight off a specific harmful germ, it needs to be "programmed" or "activated" by an antigen presenting cell (APC) that has engulfed the pathogen. This process is called priming. Once a T cell is activated, it becomes a "memory" cell, meaning it has learned how to recognize and respond to a specific pathogen.

Memory T cells remain in the body for many decades and can provide long-lasting immunity to certain infections. They can also quickly respond to future infections, allowing the immune system to mount a faster and more effective response. This is why vaccines, which introduce a harmless form of a pathogen to the body, can provide long-term immunity to certain diseases.

It's worth noting that there are different types of T cells, such as CD4+ and CD8+ T cells, which have different functions and recognize different types of pathogens. CD4+ T cells, also known as helper T cells, help activate other immune cells, such as B cells and macrophages, to fight infection. CD8+ T cells, also known as cytotoxic T cells, directly kill infected cells or tumor cells.

T cells play a vital role in protecting the body against infection and disease, and their ability to remember specific pathogens allows them to provide long lasting immunity and faster response times to future infections.

A healthy young person's body is like a well-oiled machine, with a robust immune system that is capable of fighting off infections and building a lifetime storehouse of memory T cells. T cells are a type of white blood cell that plays a central role in cell-mediated immunity, and they are essential for protecting the body against viral and bacterial infections.

When a young person is exposed to a new infection, their immune system mounts a response by producing naïve T cells, which are T cells that have not yet encountered an antigen and are not yet specific to any particular pathogen. These naïve T cells are able to recognize and respond to a wide range of antigens, and they are an important part of the body's defense against new infections.

As a person ages, however, their body produces fewer naïve T cells. This means that they have a smaller pool of T cells that are able to recognize and respond to new infections. At the same time, the body's ability to produce new T cells in response to infection is also reduced with age. This can make older people more susceptible to infections and less able to fight them off effectively.

In addition to producing fewer naïve T cells, the immune system also undergoes other changes with age. For example, the thymus gland, which is responsible for producing T cells, begins to shrink with age, leading to a decrease in the number of T cells produced. Additionally, the T cells that are produced may be less effective at recognizing and responding to antigens.

One way to combat the decline in immune function with age is to develop vaccines that are specifically designed for older people. The shingles vaccine, which is effective in older people, is an example of this. Since shingles is the reactivation of the chickenpox virus, this particular vaccine relies on existing memory T cells and does not require naïve T cells to produce a protective immune response.

Researchers are investigating ways to develop other vaccines that are adjusted for the changes that happen in an older person's immune system. For example, they are exploring the use of different adjuvants, which are substances that help to enhance the immune response, in vaccines for older people. They are also looking at ways to stimulate the production of naïve T cells in older people, such as through the use of stem cells or other therapies.

A healthy young person's body is like a well-oiled machine, with a robust immune system that is capable of fighting off infections and building a lifetime storehouse of memory T cells. However, with age, the immune system undergoes changes that can make it less effective at combating new health threats. Developing vaccines that are specifically designed for older people, such as the shingles vaccine, is one way to combat this decline in immune function. Researchers are continuing to investigate ways to develop other vaccines that are adjusted for the changes that happen in an older person's immune system, with the goal of protecting older people from infections and keeping them healthy.

Immunosenescence is a term that describes the decline of the immune system with age. It is a complex process that involves the interplay of various factors, including genetics, lifestyle, and environmental factors. The immune system is made up of both innate and adaptive components, and both of these components are affected by aging.

The innate immune system is the first line of defense against pathogens and is activated immediately upon infection. It includes physical barriers such as the skin and mucous membranes, as well as cells and proteins that can recognize and attack invading pathogens. The innate immune system is generally effective at fighting off infections in younger individuals, but its function declines with age.

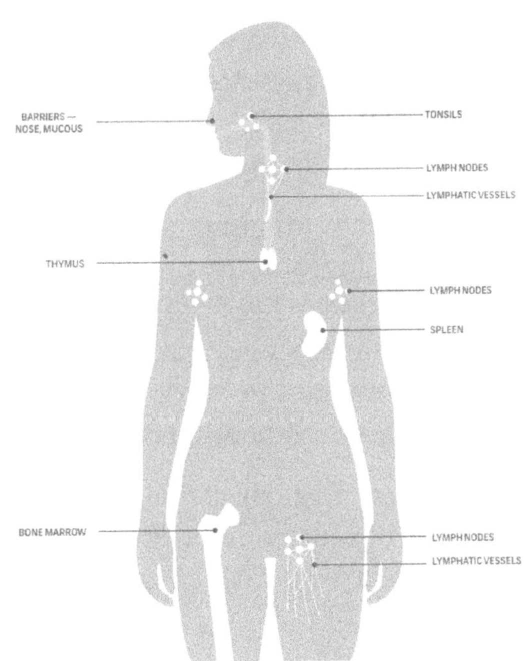

The adaptive immune system, on the other hand, is activated after the innate immune system and involves the recognition of specific pathogens by T cells and B cells. T cells, also known as T lymphocytes, are a type of white blood cell that can recognize and kill infected cells or produce chemical signals that activate other immune cells. B cells, also known as B lymphocytes, are responsible for producing antibodies, which are proteins that recognize and bind to specific pathogens, neutralizing them. The adaptive immune system is also affected by aging, as the number and diversity of T cells and B cells decrease over time, making it more difficult for the body to fight off infections.

Researchers believe that a lifetime of stress on the body, including exposure to radiation, chemicals, and certain diseases, can contribute to the decline of the immune system. Additionally, lifestyle factors such as smoking, lack of exercise, and poor diet can also play a role in immunosenescence.

Studying the immune system and how it ages is crucial in understanding how to prevent or reverse the decline in immune function. Researchers are working to identify the specific areas of the immune system that are most vulnerable to aging and to develop strategies to boost immune protection in older individuals. Some of these strategies include:

- Exercise: Regular exercise has been shown to improve immune function in older adults. Exercise can increase the production of white blood cells, which are an important part of the immune system.

- Diet: Eating a healthy diet rich in fruits, vegetables, and whole grains can help support immune function.

- Vaccines: Vaccines can help protect against specific diseases and can be especially important for older adults who may be more susceptible to infections.

- Antioxidants: Antioxidants, such as vitamins C and E, can help protect cells from damage caused by free radicals, which are unstable molecules that can damage cells and contribute to aging.

- Immunotherapy: Immunotherapy, also known as biologic therapy, is a type of treatment that uses the body's immune system to fight disease. Immunotherapy can help boost the immune system and may be effective in treating certain diseases, such as cancer and autoimmune disorders.

While these strategies may help support immune function, there is still much work to be done to understand how to reverse the decline in immune function that occurs with age. Ongoing research may shed light on new ways to boost immune protection in older individuals and improve health.

EPIGENETICS

Epigenetics is a fascinating field of study that reveals the intricate relationship between genetics and the environment. It shows how the external factors we expose ourselves to can shape the way our genes function, influencing our development, health, and aging. This new understanding challenges the long-held notion that genetics and environment are separate entities, and instead, highlights the dynamic interplay between the two.

The study of epigenetics has far-reaching implications. By understanding how environmental factors influence gene expression, researchers can identify potential targets for therapies and interventions that could prevent or even reverse the negative effects of certain diseases. For example, studies have shown that exposure to toxins, stress, or poor nutrition can lead to epigenetic changes that increase the risk of developing conditions such as cancer, diabetes, or mental health disorders.

Moreover, epigenetics can help us better understand the aging process. As we age, our bodies undergo a range of changes that can affect our health and well-being. Epigenetic changes can contribute to these changes, and research in this area may lead to new strategies for maintaining health and preventing age-related diseases.

In addition, the study of epigenetics raises important questions about the role of personal responsibility in health and disease. If our environment can shape our gene expression, then it follows that our lifestyle choices and exposure to certain environmental factors can also play a role in our health. This realization can motivate us to take a more proactive approach to our health, making choices that support healthy gene expression and reducing our exposure to harmful environmental factors.

The emerging field of epigenetics offers exciting opportunities for scientific discovery and has the potential to transform our understanding of health and disease. By exploring the dynamic interplay between genetics and the environment, we can develop new strategies for promoting healthy development, preventing disease, and improving quality of life across the lifespan.

The epigenome is a complex system that plays a crucial role in regulating gene expression and is influenced by a variety of factors, including lifestyle, environment, and genetics. Epigenetic changes can affect gene activity and may contribute to the development or exacerbation of diseases and conditions, such as age-related diseases. In some cases, these epigenetic changes can be inherited by offspring, highlighting the importance of understanding the interplay between genetics and the environment in shaping our health.

One of the key findings in recent years is that epigenetic changes can be influenced by a person's lifestyle and environment. For example, studies have shown that a person's diet, exposure to toxins, and stress levels can all impact the epigenetic marks on their DNA. These changes can affect the expression of genes involved in a variety of biological processes, including metabolism, inflammation, and cell growth.

In addition, research has found that some epigenetic changes can be inherited by offspring. This means that the environmental factors that a person is exposed to during their lifetime can have a lasting impact on the health of their children and grandchildren.

For example, studies have shown that exposure to toxins, poor nutrition, and stress can all lead to epigenetic changes that are passed down to future generations.

Understanding the epigenome and its relationship to the environment is crucial for developing effective strategies for disease prevention and treatment. By identifying the specific epigenetic changes that are associated with certain diseases, researchers may be able to develop targeted therapies that can modify these marks and improve gene expression. Additionally, by understanding how lifestyle and environmental factors impact the epigenome, individuals may be able to make changes that reduce their risk of developing certain diseases.

The epigenome is a complex system that is influenced by a variety of factors, including lifestyle, environment, and genetics. Epigenetic changes can affect gene activity and may contribute to the development or exacerbation of diseases and conditions. By understanding the interplay between the epigenome and the environment, researchers may be able to develop effective strategies for disease prevention and treatment, and individuals may be able to make informed decisions about their lifestyle and health.

Identical, maternal twins are ideal for epigenetic research because they provide a unique opportunity to study the effects of environmental and lifestyle factors on gene expression. Since they have nearly the same genetic blueprint at birth, any differences that arise between them over time are likely due to external influences. By carefully studying these changes, scientists can gain valuable insights into how environmental and lifestyle factors can shape gene expression and potentially contribute to the development of disease.

One of the key advantages of studying twins is that they allow researchers to control for genetic variables, which can help to isolate the effects of environmental and lifestyle factors. By comparing the epigenetic profiles of identical twins who have been raised in different environments or have different lifestyles, scientists can identify specific changes that are associated with specific exposures or behaviors.

For example, a study published in the journal Nature in 2019 found that identical twins who were raised apart had differences in their epigenetic profiles that were associated with differences in their environmental and lifestyle exposures. The study found that twins who were raised in different households had different patterns of gene expression related to immune function, metabolism, and brain development, among other traits.

Another study published in the journal Science in 2018 found that identical twins who had different lifestyles, such as different diets and exercise habits, had differences in their epigenetic profiles that were associated with their lifestyle choices. The study found that twins who had a healthier lifestyle had more active genes related to healthy aging and longevity, while twins who had a less healthy lifestyle had more active genes related to disease and aging.

While studying twins provides valuable insights into the effects of environmental and lifestyle factors on gene expression, it is important to note that twin studies are not without limitations. For example, twins may not always be perfectly identical, and there may be variations in their genetic makeup that are not immediately apparent. Additionally, twins may have different exposures and experiences that are not accounted for in the study.

Despite these limitations, twin studies have already led to important discoveries in the field of epigenetics and have the potential to reveal even more about the complex interplay between genes, environment, and lifestyle. As researchers continue to explore the epigenetic landscape of twins, they may uncover new insights into how we can optimize our health and well-being throughout our lives.

Epigenetics is a fascinating field that highlights the complex interplay between genetics and environmental factors. The example you provided about laboratory mice that are genetically identical but have varying lifespans due to differences in nurturing is a great illustration of how epigenetics can influence aging.

It's thought-provoking to consider how seemingly small differences in early life experiences, such as the amount of milk a mouse receives from its mother or the amount of grooming it receives, can have a profound impact on the development and aging of an organism. These experiences can shape the expression of genes involved in aging, leading to differences in lifespan.

This research has implications for understanding aging in humans as well. While humans are much more complex than mice, the study of epigenetics can provide insights into how environmental factors, such as early life experiences, nutrition, and stress levels, can influence aging and longevity.

It's also worth noting that epigenetics can be influenced by a variety of factors, including lifestyle choices and environmental exposures. This means that individuals may have some control over their epigenetic profile, and therefore, their aging process.

The study of epigenetics is a rapidly evolving field that has the potential to reveal new insights into the aging process and how it can be influenced by a variety of factors.

As research in epigenetics advances, scientists aim to address three crucial questions:

- How do modifications in the epigenome result in long-term disparities in health and aging?

- Can singular events impact the epigenome?

- If singular events can modify the epigenome, does the organism's age (or stage of development) at the time of the alteration have significance?

Understanding the answers to these questions can provide valuable insights into the mechanisms underlying health and aging, potentially leading to the development of novel therapeutic strategies for the prevention and treatment of various diseases.

STEM CELLS &

REGENERATIVE MEDICINE

Reversing age-related, chronic degeneration and restoring the body to its original health and vigor is a fascinating concept that has garnered significant attention in recent years. While the idea of regenerative medicine is still in its infancy, research on stem cells has shown promise in addressing various age-related diseases and improving health.

Stem cells are undifferentiated cells that have the ability to differentiate into specialized cell types, and they play a crucial role in the body's natural healing process. By harnessing the power of stem cells, researchers and clinicians hope to develop new treatments that can repair damaged tissues, organs, and systems, effectively reversing the effects of aging and age-related diseases.

Some of the most promising areas of research in regenerative medicine include:

- Mesenchymal stem cells: These stem cells have the ability to differentiate into various cell types, including bone, cartilage, fat, and muscle. Researchers are exploring their potential in treating conditions such as osteoarthritis, rheumatoid arthritis, and degenerative disc disease.

- Induced pluripotent stem cells: These stem cells are derived from adult cells, such as skin or blood cells, and are reprogrammed to have the ability to differentiate into any cell type in the body. Researchers are investigating their potential in treating conditions such as heart disease, diabetes, and neurodegenerative disorders.

- Tissue engineering: This involves the use of stem cells, biomaterials, and bioactive molecules to create functional tissue substitutes that can replace or repair damaged tissues. Researchers are exploring the use of tissue engineering in treating conditions such as organ failure, bone fractures, and skin wounds.

- Gene editing: Gene editing technologies, such as CRISPR/Cas9, allow for the precise modification of genes to correct genetic defects or mutations. Researchers are investigating the use of gene editing in treating genetic diseases, such as sickle cell anemia and cystic fibrosis.

While these advancements are promising, it's essential to note that regenerative medicine is still in its early stages, and significant challenges remain before these treatments can be widely available. Safety concerns, efficacy, and ethical considerations must be carefully addressed before these therapies can be brought to market.

While the idea of reversing age-related, chronic degeneration and restoring the body to its original health and vigor is exciting, it's essential to approach these developments with caution and recognize the complexity and challenges associated with regenerative medicine. Nonetheless, the potential benefits of these therapies offer hope for a healthier, more vibrant future.

Adult stem cells have the potential to be used in regenerative medicine approaches because they can be obtained from the patient's own body, they can differentiate into various cell types based on the location and other factors, and they can continue to function normally throughout an almost infinite number of cell divisions. This means that if stem cells are inserted into a damaged part of the body, they could potentially develop into area-specific cells that could restore function.

Researchers are currently exploring the use of adult stem cells in various regenerative medicine approaches, such as tissue engineering, organ transplantation, and cell-based therapies. For example, researchers are studying the use of adult stem cells to regenerate damaged heart tissue in individuals with heart disease, to repair damaged bones and cartilage in individuals with orthopedic injuries, and to restore function in individuals with neurological disorders such as Parkinson's disease.

While the use of adult stem cells in regenerative medicine approaches holds great promise, it is still a relatively new and evolving field, and there are many challenges that researchers must overcome before these therapies can be widely available. For example, researchers must develop methods to isolate and expand the stem cells, deliver them to the damaged tissue, and ensure that they differentiate into the desired cell type. Additionally, there are concerns about the safety and ethical implications of using adult stem cells, as well as the potential for tumor formation or unintended side effects.

Despite these challenges, the use of adult stem cells in regenerative medicine approaches has the potential to revolutionize the way we treat a wide range of diseases and injuries. As research in this field continues to advance, we may see the development of new therapies that can help to restore function and improve the quality of life for individuals with a variety of conditions.

Findings from early research on regenerative medicine, primarily in animal models, which show potential for stem cell treatment

- Mouse ovarian cells: In a study published in the journal Nature in 2011, researchers demonstrated that infertile mice could be restored to fertility by inserting mouse ovarian cells created from a female donor's stem cells. The study showed that the transplanted cells were able to differentiate into eggs that could be fertilized and develop into healthy pups. This study suggests that stem cells may be a promising approach for treating infertility in humans.

- Reactivating existing stem cells: Another study published in the journal Nature Communications in 2017 found that function could be restored to injured muscle tissue in mice by reactivating existing stem cells rather than transplanting new ones. The researchers used a small molecule to activate the dormant stem cells, which then differentiated into muscle fibers and replaced the damaged tissue. This study suggests that reactivating dormant stem cells may be a more effective and efficient approach than transplanting new stem cells.

- Restoring vision: In a 2010 Italian study with human participants, researchers restored vision to some people with severe burns on the outer layer of their eyes (the cornea) by using stem cells grown in the laboratory. The study involved transplanting a sheet of stem cells onto the damaged cornea, where they differentiated into healthy corneal tissue. This study suggests that stem cells may be a promising approach for treating eye injuries and diseases.

These findings from early research on regenerative medicine in animal models suggest that stem cells may be a promising approach for treating a range of diseases and injuries, including infertility, muscle damage, and eye injuries. However, more research is needed to fully understand the safety and efficacy of stem cell therapies in humans.

Researchers have been exploring alternatives to stem cells that have similar healing abilities and can be used in regenerative medicine. One such alternative is induced pluripotent cells (iPSCs), which are cells that have been reprogrammed to have the ability to differentiate into any cell type in the body, similar to stem cells.

iPSCs are created by taking adult cells, such as skin cells, and introducing specific genes that are typically expressed in stem cells. This process "reprograms" the adult cells to have the ability to differentiate into any cell type in the body, much like stem cells.

iPSCs have several advantages over stem cells. They can be generated from a patient's own cells, which eliminates the risk of rejection and the need for immunosuppressive drugs. They also have the potential to be used in a wide range of applications, including tissue engineering, drug development, and personalized medicine.

However, iPSCs are not without their challenges. They can be difficult to generate and require a significant amount of time and resources. They also have a risk of developing abnormalities or becoming cancerous, which must be carefully monitored and controlled.

Despite these challenges, researchers are making rapid progress in developing iPSCs for regenerative medicine. They are exploring ways to generate iPSCs more efficiently and safely, and are testing their use in a variety of applications, such as treating heart disease, diabetes, and neurological disorders.

iPSCs offer a promising alternative to stem cells for regenerative medicine, and researchers are continuing to explore their potential in this exciting field.

Let's explore some of these questions in more detail.

- Do older adults have enough stem cells for this type of therapy or do they need to be donated from someone else?

The number and quality of stem cells in the body decline with age, which can make it more difficult to obtain enough stem cells for therapy. However, researchers have developed methods to expand and rejuvenate stem cells in the lab, which can help overcome this limitation. Additionally, induced pluripotent stem cells (iPSCs) can be generated from an older person's skin cells, which can be used for therapy.

- Would creating stem cells from an older person's skin cells work?

Yes, it is possible to generate induced pluripotent stem cells (iPSCs) from an older person's skin cells. The process involves taking skin cells, such as fibroblasts, and introducing specific genes that are typically expressed in stem cells. This process can reprogram the skin cells into a pluripotent state, which means they have the ability to differentiate into any cell type in the body.

- Would stem cell therapy restore health and vigor to an older person or only work in a younger person?

Stem cell therapy has the potential to restore health and vigor to an older person, but its effectiveness may vary depending on the individual and the specific condition being treated. Some studies have shown that stem cell therapy can improve tissue function and reduce inflammation in older adults, but more research is needed to fully understand its potential benefits and limitations.

- How would stem cell therapy work on the cellular level—would stem cells replace the non-functioning cells or would they reactivate and repair the damaged cells?

Stem cell therapy can work in two ways: by replacing damaged or non-functioning cells, and by reactivating and repairing damaged cells. When stem cells are introduced into the body, they can differentiate into the cell type that is needed to repair the damaged tissue. They can also produce growth factors and other signaling molecules that can help to reactivate and repair damaged cells.

- Would stem cell therapy work in all areas of the body, or only in some areas?

Stem cell therapy has the potential to work in various areas of the body, including the brain, heart, liver, and pancreas. However, its effectiveness may vary depending on the specific tissue or organ being targeted. For example, stem cell therapy has shown promise in treating damaged heart tissue after a heart attack, but its effectiveness in treating neurodegenerative diseases such as Alzheimer's is still being studied.

While there are still many questions about stem cell therapy, researchers are making progress in understanding its potential benefits and limitations. It's important to note that stem cell therapy is not a one-size-fits-all solution, and its effectiveness may vary depending on the individual and the specific condition being treated.

The potential of regenerative medicine to treat degenerative diseases is vast and promising. Degenerative diseases, such as

Alzheimer's, Parkinson's, and heart disease, are characterized by the progressive loss of cellular function and tissue damage. Regenerative medicine approaches, including stem cell therapy, tissue engineering, and gene editing, offer the possibility of repairing or replacing damaged cells and tissues, potentially halting or even reversing the progression of these diseases.

For example, stem cell therapies have shown promise in treating heart disease by promoting the regeneration of healthy heart tissue and improving cardiac function. Similarly, gene editing technologies, such as CRISPR, have the potential to correct genetic mutations that underlie many degenerative diseases, while tissue engineering approaches could provide new, healthy tissue to replace damaged or diseased tissue.

While significant challenges remain, including the need for more research and development, the potential of regenerative medicine to transform the treatment of degenerative diseases is substantial. With continued advancements in technology and scientific understanding, it is likely that regenerative medicine will play an increasingly important role in the treatment of a wide range of diseases, improving the quality of life for millions of people around the world.

STRESS

In the world of genetic manipulation and life extension, the tiny nematode worm Caenorhabditis elegans, or C. elegans, stands as a pivotal model organism. This creature, with its surprisingly complex biology despite its minuscule size, has offered profound insights into the mechanisms underlying longevity. Among these discoveries was a milestone moment when scientists engineered the first C. elegans worm with an extended lifespan, a feat that heralded a cascade of revelations.

This groundbreaking accomplishment unveiled more than just an elongated existence in these creatures. The key revelation lay in the worm's newfound resistance to stress induced by heat. This resistance became an intriguing hallmark shared by a plethora of long-lived animals across various species. In essence, it suggested a fundamental correlation between longevity and the ability to withstand diverse stressors, hinting at a deeper physiological connection.

Researchers dove deeper into this phenomenon, scrutinizing the cellular and systemic responses of these genetically altered C. elegans specimens. What emerged was a compelling narrative: these long-lived organisms exhibited a robust defense mechanism against a spectrum of stressors, not just heat-induced challenges. The resilient nature of their cells and, in some instances, the entire organism itself, became a defining trait, setting them apart from their counterparts with average or shorter lifespans.

This newfound resilience painted a mosaic of possibilities for understanding the intricate interplay between longevity and stress resistance. It begged the question: could enhancing stress resistance pathways be a gateway to extending lifespan in diverse organisms, including humans? The tantalizing prospect spurred an ambitious pursuit among scientists to decipher the intricate mechanisms governing stress resistance and longevity.

As the research advanced, a universe of interconnected pathways began to unfold, intertwining molecular intricacies and cellular dynamics. Insights gleaned from these resilient C. elegans specimens sparked hope for potential interventions in humans, offering glimpses into how manipulating certain genetic or biochemical pathways might confer resilience against a myriad of stressors, potentially enhancing longevity.

Yet, amidst this burgeoning understanding, the quest for extending human lifespan remains an enigmatic puzzle, a challenge where ethical, societal, and scientific boundaries converge. Nevertheless, the saga of the genetically enhanced C. elegans worms represents a pivotal chapter in the ongoing narrative of unraveling the mysteries of longevity, offering a glimpse into a future where the resilience against life's tribulations might hold the key to an extended existence.

A compelling shift has emerged in the selection of animal models for studying successful aging. Biologists, aiming to decode the intricate mechanisms governing longevity, have embarked on a novel path—one that revolves around scrutinizing stress resistance as a pivotal criterion in their selection process.

This innovative approach entails the meticulous evaluation of stress resistance in young animals, serving as a litmus test to identify specimens that exhibit remarkable resilience in the face of biological stressors. Instead of adopting a broad-spectrum analysis, researchers opt to selectively study only those animals showcasing high levels of stress resistance. This strategic pivot rests on a foundational hypothesis that underpins their quest: a direct and intrinsic correlation between the ability to combat biological stressors and an extended lifespan.

By subjecting young animals to a battery of stress-inducing challenges, scientists can discern nuances in their responses. Those displaying heightened resilience become the focal point of intense scrutiny, representing a subset with potential genetic, cellular, or systemic traits that contribute to their ability to withstand stressors.

This paradigm shift in animal model selection reflects an acknowledgment of the common thread linking longevity to the capacity to navigate and withstand diverse stressors. The hypothesis driving this selective approach posits that organisms equipped with robust stress resistance mechanisms may inherently possess underlying traits conducive to a prolonged and healthier lifespan.

This meticulous curation of animal models based on stress resistance represents a departure from traditional methodologies. It marks a deliberate shift toward a more focused exploration, aiming to unravel the intricate tapestry of biological mechanisms that govern both stress resilience and longevity.

However, this approach, while promising, navigates through uncharted territories, presenting challenges and uncertainties. The complexity of aging and the multitude of factors influencing longevity necessitate comprehensive and interdisciplinary investigations.

The quest to untangle the relationship between stress resistance and longevity remains an ongoing saga in the scientific pursuit of understanding the fundamental principles governing aging. The strategic shift in animal model selection heralds a new chapter in this saga, one that holds the promise of unlocking profound insights into the biology of successful aging.

The logic behind this is that the ability to withstand stresses and assaults could be indicative of robust systems and processes that may also slow the aging process. For example, animals with superior antioxidant defenses, DNA repair mechanisms, or cardiovascular function may not only be better equipped to handle exposures to things like toxins, radiation, or temperature swings. But they may also suffer less cumulative damage over time, preserving tissue, cell, and molecular function.

Ongoing studies will help determine if this hypothesis holds true and if selecting for high early-life stress resistance allows researchers to essentially identify "super-agers" amongst animal populations that are predisposed for slower aging and extended healthy lifespans. Of course, stress resistance could end up having little correlation with longevity once animals are studied across their full life cycles. But if a definitive connection is made, it opens some intriguing possibilities. Tests for stress resistance could guide anti-aging research towards those animal models that represent the extremes of robust systems and naturally delayed aging. This could accelerate insights into the biology of resisting age-related decline.

Here are a few key points to discuss regarding that statement on studying stress resistance and longevity:

 - Rationale behind the approach

- The rationale seems to be that animals with higher innate stress resistance may have cellular pathways, protective mechanisms, etc. that help preserve function and survival. Studying these processes in stress-resistant models could shed light on longevity mechanisms.

- Types of stress resistance

- The statement is vague, but researchers could look at oxidative stress, DNA damage, proteotoxic stress, thermal stress, hypoxic stress, etc. The type studied matters in terms of relating it to aging pathways.

- Potential linkages

- Increased stress resistance could indicate better protein homeostasis, mitochondrial functioning, apoptosis regulation, etc. These molecular-level protections could preserve cell viability for longer as organisms age.

- Cause and effect

- As the statement notes, ongoing research is still needed to determine if stress resistance actually directly causes/drives increased longevity or the two are only correlated. There may or may not be a causal relationship.

- Study design considerations

- Researchers would need to tightly control variables like the age testing is done for stress resistance and use sufficiently large sample sizes. The metrics for "high stress resistance" would also require clear threshold criteria.

- Translation to other species

- Findings linking stress resistance and longevity mechanisms may or may not be generalizable to other species, including humans. Further validation would be necessary.

While a promising research approach, the direct linkage between early-life stress resistance and lifespan remains to be fully scientifically validated and explored as an aging model. But the approach could yield valuable insights if carefully executed.

Scientists are also investigating if mental anguish can impact the aging process. Specifically, a number of investigations focused on if stress can accelerate telomere shortening. Telomeres are protective caps on chromosomes that gradually erode with cell division and age. One study observed that women parenting kids with severe, lasting illness displayed quicker telomere shrinkage versus other moms. Additional work saw that people nursing those with Alzheimer's also had hastened telomere diminishing. So difficult caregiving and emotional strains may potentially quicken biological aging at a cellular level.

The proposed explanation is that strain inflicted by mental hardship may quicken telomere attrition through biochemical pathways triggered by stress. These could amplify cell proliferation, inflammation, and oxidative damage. Though more research is mandatory to reveal the precise systems connecting perceived stress to genetic aging markers like telomeres. Future studies also need to follow subjects over longer periods to correlate accelerated telomere shortening with impacts on health and lifespan. Only then could scientists conclude whether stress-related telomere erosion meaningfully impairs whole body vigor and longevity rather than just altering a cellular metric.

Initial findings raise the prospect that emotional anxiety could potentially accelerate aging by speeding a key genetic marker. But more rigorous longitudinal work is necessary to both demonstrate robust links between psychological distress and quicker telomere shortening as well as confirm that tighter telomere length truly signals impaired health and function later in life rather than essentially acting as a clock of aging. The clinical ramifications are presently indefinite without further mechanistic explanations and connecting accelerated telomere erosion to tangible physical decline over time. But this remains an intriguing avenue of aging research deserving expanded attention if scientifically validated.

CHANGING OLDER ADULTS' IMMUNITY

Key points to discuss on that passage about immunosenescence and aging:

- Declining immune function with age clearly raises vulnerability to infection and likely contributes to issues like pneumonia which are life-threatening illnesses in the elderly. So there are definite health costs.

- However, overactive inflammatory immune responses also seem to play a role in autoimmune disorders, atherosclerosis, Alzheimer's, arthritis and other age-related diseases. Dampening these could help reduce such conditions.

- It's a complex balancing act - the aged immune system walks a fine line between being underactive (risking infection) or overreacting (driving inflammation). There may be an optimal level that avoids both extremes.

- The phenomenon of immunosenescence itself may have evolved protective roles. As the body ages, scaling back certain immune functions could be adaptive and prevent issues like autoimmunity or chronic tissue damage from inflammation.

- More research is still needed to unravel this complexity before interventions would be advisable. Immunomodulation could backfire and cause harm if the context-dependent functions of immunosenescence are not yet fully understood.

- Both animal and human studies looking at health/lifespan impacts of boosting or suppressing specific immune pathways in late life could reveal more about the pros/cons and help guide appropriate interventions. The aging immune system likely needs to be carefully balanced.

Substantial gaps in knowledge around immunosenescence remain before its multi-faceted effects are clear and it can be safely modulated to support healthier aging. A nuanced approach seeing it as neither universally good or bad is likely warranted based on its Janus-like mix of protective and harmful impacts that may be difficult to separate.

LAST WORDS

The age-old question surrounding the causes of aging has captivated the minds of medical researchers, philosophers, anthropologists, and the wider public across centuries. It's an enigmatic journey that this booklet merely brushes upon, offering a glimpse into the complex science behind aging.

In its essence, aging perplexes us all—we live it each day, progressing inevitably with time. While many of us might not unravel the microscopic intricacies at play, the fascination remains palpable. A young girl marvels at her vibrant grandmother, perhaps unaware that the older woman's vitality hints at potential experiences during her own golden years. Similarly, a middle-aged man may recognize the benefits of healthy choices and an active lifestyle without delving into the depths of metabolism and biological stress.

Aging, intrinsically intertwined with life, remains an integral part of our existence. The maturing field of gerontology promises ongoing revelations about the intricate internal transformations from childhood to the elder years. Surprisingly, experiments involving seemingly disparate animal models—yeast, fruit flies, worms, and mice—offer profound insights into the aging process, unveiling facets that may eventually translate into critical interventions for humans, be they clinical, pharmacological, or behavioral.

So, what lies ahead in the realm of aging biology research? The quest for a modern-day fountain of youth may remain a myth, an elusive elixir promising a return to youthfulness. Yet, research holds promise in paving the way toward a healthier, elongated life.

This prospect potentially offers the gift of more moments with loved ones, the chance to greet great-great-grandchildren, and the opportunity to savor a wealth of life experiences.

The future, steeped in the ongoing discoveries of gerontology, hints at a path illuminated by scientific insights. It's not about turning back the clock, but about enhancing the quality and duration of life. This journey, woven with the threads of scientific inquiry and human aspiration, may usher in an era where the twilight years offer not just wisdom but a vibrant tapestry of experiences, enriched by the advances in our understanding of aging.

GLOSSARY

Aging: The changes that occur throughout the course of life. Gerontologists study what distinguishes normal aging from disease.

Antioxidants: Compounds that counteract free radicals and may protect cells from damage. Examples are enzymes like superoxide dismutase (SOD) and nutrients like vitamin C.

Calorie restriction: Reducing calorie intake while maintaining nutrients. Studied in animals to explore impacts on aging.

Cell senescence: When a cell permanently stops dividing. May contribute to aging.

Centenarian: A person who has lived over 100 years. Studies of centenarians look for factors related to longevity.

DNA: Deoxyribonucleic acid that contains our genetic code. Accumulated DNA damage may contribute to aging.

Free radicals: Unstable molecules that can damage cells. Formed naturally when food is metabolized but held in check by antioxidants.

Genes: Segments of DNA that code for proteins. Researchers search for longevity genes associated with long lifespan.

Immunosenescence: Age-related decline in immune function. Studied for its role in disease risk.

Metabolism: Life-sustaining chemical processes that also create cell damage from use of oxygen and nutrients.

Mitochondria: Cell components that produce energy. May accumulate damage with age.

Oxidation: Cell damage from free radicals that unfolds through a chain reaction.

Proteins: Molecules produced by genes essential for biological functions. Examples are antioxidants SOD and catalase.

Stem cells: Cells that can divide, become many cell types, and self-renew. Studied for potential regenerative abilities.

Telomeres: Protective caps on chromosomes that shorten with cell division. Extreme shortening triggers senescence.